Lake City Jr. Academy
Library

Lake City Jr. Academy
LIBRARY

Lake City Jr. Academy
Library

WRITTEN BY
DIANE COSTA DE BEAUREGARD, CLARISSE DENIAU,
MARIE FARRÉ, ANNE DE HENNING, MAURICE KRAFFT,
LAURENCE OTTENHEIMER-MAQUET, JEAN-PIERRE VERDET

COVER DESIGN BY
STEPHANIE BLUMENTHAL

TRANSLATED AND ADAPTED BY
ROBERT NEUMILLER

PUBLISHED BY CREATIVE EDUCATION
123 South Broad Street, Mankato, MN 56001
Creative Education is an imprint of The Creative Company

© 1990 by Editions Gallimard
English text © 2001 Creative Education
International copyrights reserved in all countries.
No part of this book may be reproduced in any form
without written permission from the publisher.

Library of Congress Cataloging-in-Publication Data
[Notre planète dans l'univers. English]
Planet Earth / by Diane Costa de Beauregard et al.
(Creative Discoveries)
Includes index.
Summary: Describes the Earth's seasons, climate, weather, oceans, volcanoes,
and other features, and explains how the solar system affects them.
ISBN: 0-88682-953-4
1. Earth—Juvenile literature.   2. Astronomy—Juvenile literature.
3. Geophysics—Juvenile literature.   [1. Earth.   2. Astronomy.]
I. Deniau, Clarisse.   II. Farré, Marie.   III. Title.   IV. Series.
QB631.4.C67   1999
550—dc21   98-7124

First edition

2 4 6 8 9 7 5 3 1

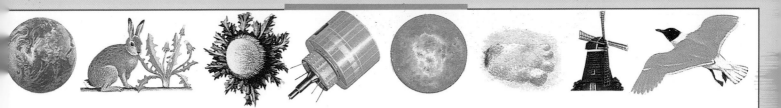

PLANET EARTH

CONTENTS

THE SOLAR SYSTEM	4
THE SUN AND ITS PLANETS	5
THE MOON	6
COMETS AND METEORS	10
WHAT ARE THE STARS?	12
THE GALAXIES; OUR GALAXY, THE MILKY WAY	16
OUR SUN AND OUR EARTH	18
WHY ARE THERE DAYS AND NIGHTS?	18
WHY ARE THERE SEASONS?	20
THE POLES, THE EQUATOR, AND THE TROPICS	22
WHY ARE THERE CLIMATES?	24
AROUND THE EARTH: AIR, WIND, AND WATER	26
RAIN, CLOUDS, AND THE WATER CYCLE	30
THUNDERSTORMS AND RAINBOWS	32
SNOW AND ICE	34
THE WEATHER	36
THE LIFE OF THE EARTH	38
OUR BLUE PLANET: THE OCEANS	38
TIDES, CURRENTS, AND WAVES	41
ICE	42
COASTLINES	43
THE EARTH'S MOVEMENTS	44
MOUNTAINS	44
VOLCANOES, LIVING MOUNTAINS	48
EARTHQUAKES	52
THE SHAPING OF THE EARTH	54
CAVES, LAKES, AND RIVERS	54
EROSION	56
A MAP OF THE EARTH TODAY	58
EXPLORE AND PLAY	61
GLOSSARY	70
INDEX	74

CREATIVE EDUCATION

The Earth circles the sun.

Without the sun there would be no life on Earth. The sun gives out warmth and light. It makes the plants grow, and people and animals in turn depend on the plants. Without the sun, the Earth would be a barren, icy desert. Yet the sun is an ordinary star, only one among billions in our galaxy. It just happens to be the star at the center of our solar system, and Earth is one of its nine orbiting planets.

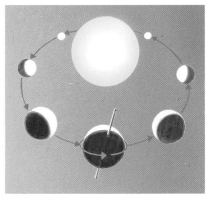

Earth takes a year to travel around the sun. It is also spinning like a top, making one complete rotation every 24 hours.

Early ideas
It was once thought that our planet was fixed at the center of the universe, and that the stars traveled around it. Although the ground seems solid and still, Earth is spinning and hurtling through space at 67,000 mph (107,803 km/h). The idea that the Earth revolves around the sun was put forth by the astronomer Nicholas Copernicus more than 450 years ago.

How does the Earth travel in space?
Every morning the sun rises in the east. It travels across the sky, and sets each evening in the west. In fact, it is the Earth itself that moves, not the sun. Earth completes one spin, or rotation, each day. At the same time, over the course of one year and six hours, the Earth travels around the sun. The extra hours give us a leap year every fourth year.

The planets of the solar system
Earth is not the only planet orbiting the sun. Starting closest to the sun, the planets are as follows: Mercury, Venus, Earth, Mars, Jupiter, Saturn, Uranus, Neptune, and Pluto. Other traveling companions are our moon, the moons of other planets, comets, and asteroids.

The moon, the Earth, and other planets do not give off any light or heat of their own. They shine only because they reflect light from the sun.

Earth is part of the Solar System.

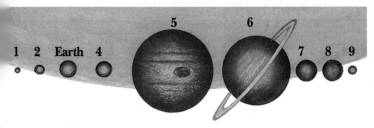

The nearest planet to the sun is Mercury (1). Earth is third from the sun. The picture shows the planets' sizes in relation to each other, but not the distances between them.

Mercury (1), Mars (4), and Pluto (9) are small like Earth and have a solid rocky surface. Jupiter (5), Saturn (6), Uranus (7), and Neptune (8) are large planets surrounded by thick atmospheres of gas. Other space travelers—much smaller than the planets—are comets and asteroids. Comets are frozen balls of gas in regular orbits. Asteroids come from the many small fragments crowded between Mars and Jupiter. They may be the remains of exploded planets or old comets.

The sun is a million times larger than Earth. But the sun is only a small star compared with other stars in the universe.

Venus (2) is somewhat smaller than Earth, but with a temperature of more than 800 degrees Fahrenheit (427°C), nothing can live there. Earth, farther away from the sun, has a more comfortable temperature.

Venus is nicknamed the Morning Star and the Evening Star. At dawn, Venus remains visible after the stars have faded. At nightfall, Venus, along with Mercury, appears first.

The rings of Saturn (6) are not solid; they are made up of frozen gases, ice, and rock fragments ranging in size from smaller than a grain of sand to as large as a house. The rings reach out as far as 84,650 miles (136,200 km) from the center of Saturn. Many of these rings are only 16 feet (5 m) thick. Astronomers once thought that there were only a few rings around Saturn, but instruments on the Voyager 2 satellite counted more than 100,000 rings.

The moon is Earth's companion in space.

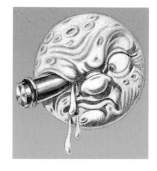

Throughout the ages, people have been fascinated by the moon. Some have seen a human or animal face in the patterns on its surface. If you promise somebody the moon, you are promising something quite impossible. It seemed impossible that men could actually walk on the moon—until July 20, 1969, when astronauts Neil Armstrong and Edwin "Buzz" Aldrin stepped out of Apollo 11 onto that dry and dusty surface.

Earth is the planet closest to the sun to have a moon. With a diameter of 2,160 miles (3,456 km), ours is one of the largest moons in the solar system.

Without the sun, we would never see the moon.

On July 20, 1969, two American astronauts, Neil Armstrong and Edwin "Buzz" Aldrin, landed on the moon in Apollo 11, a special lunar craft. They brought back to Earth the first samples of lunar soil and rocks for scientific examination.

The moon shines because it reflects the light it receives from the sun. At full moon, dark markings become visible on the moon's surface. These are the lunar seas, or maria. There is no water on the moon, although scientists believe ice may exist in the deep craters. The maria are vast plains covered with a dark-colored dust. The pale marks are the lunar mountains, the tallest at 25,688 feet (7,800 m) high.

Through a pair of good binoculars, these lunar mountains are clearly visible. The craters of all sizes that cover the moon are also visible.

Photographs taken from spaceships show the other, hidden face of the moon. This side is invisible from Earth. In the time it takes the moon to spin once, it also makes one full journey around Earth, so it always points the same face towards us.

The moon has been bombarded by meteorites for billions of years, creating the craters on its surface. Moon rocks brought back by the astronauts have been tested by scientists, and the results show that the moon and the Earth had a common origin. They are both about 4.5 billion years old. Moon rocks turned out to be similar to some of the rocks found on Earth.

During a full moon it becomes light enough to see at night.

The Earth and the moon play hide and seek.

If the moon shone with its own light, we could see it all the time. However, because the moon and the Earth are both lit by the sun, what we see of the moon depends on its position in relation to the Earth and the sun. These regular changes in the way the moon looks in the sky are called the phases of the moon. The moon's phases helped people in the past keep track of time.

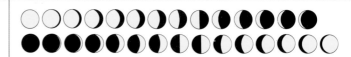

The moon takes 27.3 days, or one lunar month, to complete its phases. Sometimes, the moon is full twice in one calendar month. This is a called a blue moon and happens only once every few years.

After the new moon is obscured from view, the new moon appears in the sky as a thin crescent. It grows wider: from a quarter, to a half, and to a full circle called a full moon. Then, night by night, it becomes smaller and thinner again.

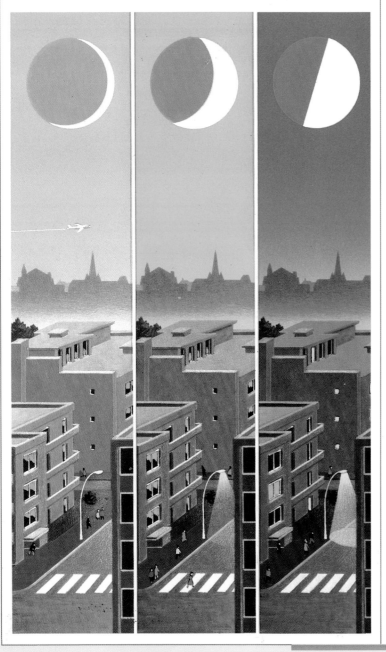

The moon waxes and wanes each month.

The moon is between the sun and the Earth at the time of new moon. The side being lit by the sun faces away from the Earth, and we see none of it. The thin crescent following the new moon is the edge of the sunlit side. As the days pass, the moon's position changes and more of it becomes visible. A full moon occurs when the moon reaches the side of the Earth opposite from the sun. The moon sets as the sun rises.

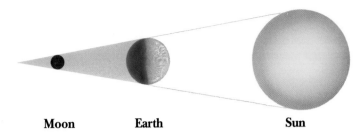

Moon   Earth   Sun

The positions of the moon, the Earth, and the sun at the time of an eclipse of the moon

Eventually the moon develops into a pale half moon in the morning sky. The part lit by the sun becomes smaller and smaller until it is just a thin sliver. The lunar month begins again, and in a few days the moon once more appears as a crescent in the evening sky. The weeks and months of our calendar are calculated from the phases of the moon.

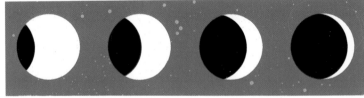

The moon slowly moves into Earth's shadow. At total eclipse, the moon becomes invisible from Earth for a few moments.

Eclipses of the moon

Like everything else touched by the sun's light, the moon and the Earth cast shadows. Sometimes in their game of hide-and-seek, the moon passes through Earth's shadow, causing what's called a lunar eclipse. At first only a small part of the moon is dark, but in a few minutes Earth's shadow covers it entirely. One total eclipse of the moon usually occurs each year.

Lake City Jr. Academy
Library

Sometimes strange lights flash across the sky.

The flash of a comet disturbs the predictable patterns of the night sky. A comet is a frozen ball of gas with a long, shining tail produced by the heat of the sun. A comet's tail is only lit up as the comet travels close to the sun. Because of the strength of the sun's radiation, the comet's tail always points away from the sun.

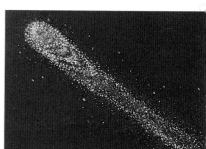

Halley's comet, with its brilliant tail of luminous dust and gas, appears about every 76 years.

In times gone by, people blamed comets for losses at war or for natural disasters. Until the middle of the 16th century, people thought comets were burning vapors rising from swamps.

In 1758, a comet passed across the skies and was given the name Halley's comet in honor of the British astronomer, Edmond Halley, who had predicted its appearance.

Meteorites—fireworks from heaven

Some artificial satellites have solar panels to convert the sun's energy into electricity to power their instruments.

Artificial satellites

We send artificial satellites into space to orbit the Earth for military purposes, communications, such as television broadcasting, and weather monitoring.

Have you ever seen shooting stars? These are meteors, many only as large as a grain of sand. Sometimes pieces of cosmic matter pass close to Earth and are pulled into its atmosphere. They travel fast, becoming so hot that they turn to gas when in contact with the air. This hot gas shines in a bright trail across the sky.

In 1908, a meteorite or piece of comet destroyed a 1,500 square mile (2,400 sq km) area in the forests of Siberia.

Most meteorites burn up completely, and only the largest ever land on Earth. The biggest known meteorite weighed 132,000 pounds (59,800 kg).

Meteor Crater in Arizona, 3,936 feet (1,200 m) in diameter, was hollowed out by a giant meteorite.

The birth of the stars

On a clear night, about 2,500 stars are visible in the sky. Many stars form recognizable groups called constellations.

Stars are being born every minute somewhere in the universe.

Often stars are born in clusters, as huge clouds of hydrogen. A powerful telescope shows stars being born in the Trifid Nebula.

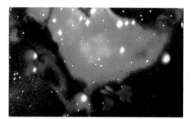 

A cloud of gas and dust starts to collapse. As this happens, a star begins to take shape.

The more the cloud collapses, the hotter the gas becomes. Eventually the gas gets so hot that it begins to shine with light. A star is born.

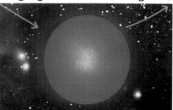 

When the star has burned its hydrogen, it grows into a red giant, and then shrinks to a white dwarf.

How do stars live and die? Stars are enormous balls of hot gas. They are so hot that they shine brightly, and in a unique way, by transforming their hydrogen into another gas called helium. The change gives out light energy and may go on for millions or billions of years.

Eventually stars burn out. In the final stage of a star's life, it may explode in an event called a supernova. During a supernova, a star becomes many times brighter, then slowly fades to less than its original brightness. In 1987, a supernova appeared in the constellation Dorado in the southern sky. It was the first supernova visible to the naked eye since 1604.

Constellations visible in the Northern Hemisphere in the southern summer sky:

1. Lyra
2. Hercules
3. Ophiuchus
4. Corona Borealis
5. Auriga
6. Virgo
7. Libra
8. Sagittarius
9. Capricorn
10. Aquarius
11. Pisces
12. Pegasus
13. Delphinus
14. Aquila
15. Cygnus
16. Sepens
17. Scorpio

The stars have long been maps for travelers.

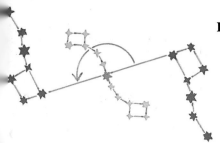

The Greater Bear and the Lesser Bear turn around the North Star each night. The North Star stays in the same spot in the sky.

The stars seem to be arranged in patterns across the sky. These are the constellations. The Big Dipper, made up of seven stars, is part of the constellation Ursa Major, or the Greater Bear. The Little Dipper is part of the constellation Ursa Minor, or the Lesser Bear. The North Star, also called Polaris, is the fixed point in the sky around which all the other stars seem to turn. Travelers in the Northern Hemisphere have used the North Star to guide them for thousands of years.

These are some of the constellations visible in the northern sky during winter in the Northern Hemisphere:

1. Orion
2. Taurus
3. Perseus
4. Aries
5. Auriga
6. Andromeda
7. Pegasus
8. Pisces
9. Aquarius
10. Eridanus

Taurus   Leo   Ursa Major

During the year, Earth passes by the 12 constellations of the Zodiac. Astrologers believe these constellations have influence over our lives.

The 12 constellations of the Zodiac:

Gemini	Pisces	Sagittarius	Virgo
Taurus	Aquarius	Scorpio	Leo
Aries	Capricorn	Libra	Cancer

Our star, the sun

People once thought the sun was a fireball burning in the sky like a great lump of coal. A lump of coal the size of the sun would have burned out after only 7,000 years. The sun has been shining much longer than that.

Like all the other stars, the sun is a ball of hot gas. Its surface burns at 11,000° Fahrenheit (6,000°C). The temperature at its center reaches 20 million degrees Fahrenheit (11.1 million degrees C).

The upper drawing shows how people 400 years ago believed the sun looked. The lower picture shows the sun as it looks when viewed through a telescope.

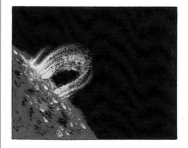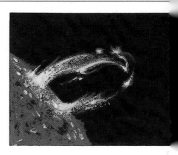

Called solar flares, giant loops of hydrogen gas are thrown off the boiling surface of the sun.

The sun is an enormous source of energy, and is essentially a huge hydrogen bomb that has been exploding and giving off energy for billions of years. Scientists expect it will carry on for many billions of years. If we imagined the sun as a person, it would be middle-aged. After the sun uses up all its hydrogen energy, it will become a cold dwarf star and will no longer shine.

For now the sun shines continuously. Some of the sun's energy hits the Earth, warming us and the plants and animals that live here. The sun's distance from Earth allows it to warm us without burning us up. The sun's energy also reaches us as light. This light is powerful enough to burn our eyes, making it dangerous to stare directly at the sun.

Earth receives the sun's energy even through thick clouds or fog. These only block a little of the light and heat. Although Earth's temperature may seem cold at times, it would be far colder if the sun no longer existed. In fact, without the sun, life on Earth would be impossible.

The sun hides behind the moon during a solar eclipse.

How can the moon hide the sun? It seems impossible, because the moon is much smaller than the sun. Since the moon is so much closer to Earth, it looks the same size as the sun. The Earth takes a year to travel once around the sun, but the moon goes around the Earth in one month. About once a year, the moon passes directly in front of the sun, hiding it from a portion of the Earth. When this happens, it becomes as dark as night during the daytime. This is called a solar eclipse.

On the parts of Earth directly in line with the sun and moon, the effect of the eclipse looks like the picture on the right.

The next solar eclipse visible in the United States will be in 2017.

The Milky Way is like an island . . .

The universe contains 40 billion to 50 billion galaxies. Although our galaxy seems huge to us, it is smaller than others, such as Andromeda, the galaxy closest to our own.

This is what the Milky Way galaxy looks like from the side.

Our solar system is only a small part of one galaxy in the universe. Our galaxy, called the Milky Way, alone contains 200 billion stars. On a clear moonless night, the Milky Way looks like a hazy silvery band, but it is made up of many stars. It has arms that float out in delicate spirals from its center. From the side it looks like a giant flattened wheel. The Milky Way is rotating, carrying us with it, on a journey around the universe. Each complete rotation the Milky Way makes takes 220 million years to complete.

If we could travel beyond our galaxy and look straight at it, it would look much like the spiral galaxy in the picture.

The small red spot represents our solar system within the Milky Way.

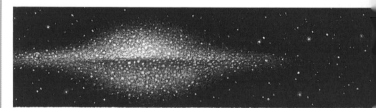

Not all galaxies are spiral in shape like ours. Others look like elliptical, or oval-shaped, blobs of light. But all galaxies contain billions of stars.

There are other areas of misty light in the night sky that are difficult to see. They are called nebulae—which means indistinct and hazy. Some are galaxies and star systems. Others are large clouds of gas and dust, like the Trifid Nebula. Some nebulae emit their own light while others reflect the light of stars.

. . . among the billions of galaxies in the universe.

The Earth and the sun are on the edge of the Milky Way galaxy. Earth's only natural satellite is the moon, but many artificial satellites now orbit the Earth. Once launched, these satellites stay in orbit around the Earth. Gravity is the force that holds us and all things around us on the Earth; it also keeps the moon and all the other satellites in orbit.

Advances in technology have produced more and more powerful Earth-based telescopes, making it possible for astronomers to see stars that were once invisible. There are even telescopes, like the Hubble Space Telescope, orbiting the Earth as they look at the stars.

The largest optical telescope in the world is in the nation of Georgia.

The sun's daily journey across the sky

Day follows night, winter follows summer, and the years come and go in an endless succession. Our lives are regulated by the movement of the Earth around the sun.

Humans and other animals are accustomed to the way the solar system works. After a day, we need rest and sleep, which we usually take at night. These changes govern the rhythm of our daily lives, and without this rhythm, life would be exhausting.

The sun rises each day in the east.

The sun looks as though it follows a great curve across the sky each day. In the mornings the sun rises in the east and, until midday, climbs in the sky. Then it begins to drop until, at the end of the day, it sets in the west. As the sun disappears, the sky grows darker and it becomes night.

Because the Earth is round, only one side at a time is lit by the sun. The half of the Earth that faces the sun has daytime, while on the other side, it is night. Earth itself spins like a top, so all parts are lit by the sun's light in turn. When the sun sets, our homes pass from day into night as they move out of the sun's light into the Earth's shadow. When the sun rises again, our homes move out of the darkness and into the warm light of the sun once more. But the sun doesn't move around the Earth. Earth travels in a great circle around the sun. Our journey around the sun takes one year to complete.

Earth makes one complete rotation on its axis every 24 hours. This length of time is called a day.

Different countries have different times. At breakfast time in England, it is late at night in the United States.

The continuous cycle of day and night

During the day the valley is in sunshine. Little by little, as the sun sinks lower in the sky, the shadows grow longer.

Eventually, near sunset, only the mountain tops are in sunlight.

As the sun sinks below the horizon, the moon and the stars are the only sources of light. People in high-flying aircraft are the last to see the sun.

The four seasons

Earth makes an immense journey around the sun—about 587 million miles (939.2 million km) each year.

As the Earth travels around the sun, we experience different seasons. The Earth's surface is curved, so the sun's rays spread out when they reach it. In summer, our hemisphere is tilted toward the sun; the rays strike us more directly, making the weather warmer. In winter, the Northern Hemisphere tilts away from the sun, and it is summer in the Southern Hemisphere.

At the spring equinox, March 21, the sun takes 12 hours to cross the sky from dawn to dusk. Then as we tilt toward the sun, it rises earlier and sets later in the evenings. As daylight gets longer, our hemisphere gains more warmth.

The days are long in summer, lasting about 16 hours by the end of June. In the Northern Hemisphere, June 21, called the summer solstice, is the longest day of the year.

Earth sits at an angle, so the sun makes a slightly different journey across the sky each day of the year.

Spring, summer, autumn, and winter

At midday, the sun reaches its highest point in the sky. From the summer solstice onward, the days get shorter. At the autumn equinox, September 23, day and night are again the same length.

Because the sun's path across the sky gets shorter, we receive less sunshine each day and the air becomes colder. The shortest day of the year is the winter solstice, December 21. The sun sets at this time of year by late afternoon.

In the temperate climate of much of North America, each of the four seasons is different and distinct.

The sun is low in the sky, while the weather is cold in the winter.

Day by day the sun climbs a little higher in the sky. Spring finally arrives.

Summer has arrived when the sun is at its highest point in the sky.

The sun's path gets closer to the horizon. The days become shorter and cooler in autumn.

The seasons at the poles and the equator

Because of the Earth's tilt, sunlight and heat fall unequally on it. As the Earth orbits the sun, different parts of the Earth receive more or less heat and light at different times of the year. This gives us seasons. Near the equator, there are few distinct changes in the seasons, unlike the temperate zone of North America.

The equator never tilts away from the sun. Night and day are almost the same length at the equator, and the weather is warm year round.

The poles during summer are only slightly tilted toward the sun, so the sun doesn't set but stays low in the sky. The sun's rays here must pass through a greater amount of atmosphere than at the equator, so they lose much of their heat. This is why it's cold year round.

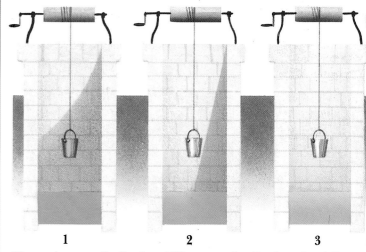

The sun's rays strike Earth at different angles. See how they fall at midday in a deep well in northern Norway (1), in southern England (2), and at the equator (3), where even the deepest well is lit right to the bottom.

Because of Earth's tilt, the sun at the poles doesn't rise in winter, leaving the areas completely dark for six months.

The North and South Poles have opposite seasons. When the North Pole basks in the midnight sun that never sets, the South Pole has its period of long winter darkness and extreme cold.

Auroras illuminate the polar night sky.

The skies of the Arctic and Antarctic are sometimes lit up with brilliant colors.

A spectacular scene
In the Arctic, this phenomenon is called the aurora borealis, or northern lights. In Antarctica, it is called the aurora australis, or southern lights. Green, purple, and gold, the lights look like shimmering curtains or giant, colored waterfalls, depending on where they appear in the sky. These splendid light displays come from the sun, which occasionally sends out a flood of charged particles, called the solar wind. Earth's magnetism traps these charged particles at the poles, where they shine brightly to create these beautiful effects. Auroras appear most often during the spring and autumn equinoxes. They are usually only visible near the poles, but the northern lights have been seen as far south as Guatemala.

Deserts of baking sand and freezing ice

Earth's climate is varied. In the United States, we live halfway between the cold, dry North Pole and the hot, humid equator. The poles are freezing deserts, where ice and snow cover the land year round. There are few animals or plants at the poles, and the temperature may fall below −60 degrees Fahrenheit (−50°C).

The parched deserts of the great continents are large regions where sometimes rain will not fall for years at a time. Rain clouds cannot develop because the ground and air are so dry. The land is scorched by day and chilled at night, so only the hardiest plants and animals can live there.

Tropical rain forests around the Earth

Long wet seasons with torrential rains every day of the year help nourish the rain forests. Dense, leafy trees grow tall above abundant shrubs and undergrowth. A wide variety of plants and animals lives in the rain forests. The Amazon Basin alone may have around 10 million species of animals, and the majority of those are insects.

The majority of North America has a temperate climate, with either rain or snow falling during the year. Four distinct seasons produce a range of different weather patterns in North America: from cold, snowy winters to hot, rainy summers. But the weather is never as extreme as at the equator or the poles.

A cloak of air surrounds the Earth.

Though we cannot see or grasp it, air surrounds us. It blows against us when it's windy and sweeps the clouds across the sky. Air is invisible, but extremely important. All living things need air to survive. Air contains oxygen, which keeps people and animals alive. Plants depend on carbon dioxide and nitrogen, other gases that make up air.

Air can hold up and support such things as airplanes, birds, hang gliders and kites.

Run fast with your arms and hands stretched out wide and feel the air pushing against you.

The layer of air surrounding Earth is called the atmosphere. Earth's gravity holds the atmosphere in place. Down at the level of the sea, the atmosphere is its thickest. Its weight is called atmospheric pressure. Because atmospheric pressure is always present, we do not notice it. But 20,000 feet (6,000 m) up, the pressure is only half of its sea-level value. Pressure becomes less and less at even greater heights, until it reaches zero out in space. This is called a vacuum.

Astronauts in spaceships travel through this nothingness and float because of the lack of gravity. If they left the protection of their ship without special suits on, their bodies would explode. Mountain climbers use oxygen masks on very high mountains because the air is thinner and breathing becomes more difficult. Without extra oxygen they would not be able to sleep, keep warm, or climb. At the top of Mount Everest, the air pressure is only about one-third of what it is at sea level.

The atmosphere acts like a filter to protect us from the sun's most damaging rays. The sun gives off an enormous amount of heat and light, but Earth receives very little of it. Most of the sun's rays bounce off the Earth's atmosphere into space.

The effects of the atmosphere on Earth

The sun's rays penetrate the atmosphere and heat up Earth's surface. Fortunately, during the night much of this ground heat escapes again. If it didn't, it would become so hot that life on Earth would be impossible.

The atmosphere also acts like a glass roof. By day it filters the sun's rays. By night it insulates, holding in some of the heat so the Earth does not cool down too much.

The air scatters the light from the sun. Without air, the sun and the stars would shine at midday against a black sky. The air molecules in the atmosphere reflect light back to us, which we see as the color blue. Once the sun has set, its rays no longer light up the air. In space there are few molecules to reflect light, so it looks black.

The sun rises. Sunlight is reflected off the water in the damp air, turning the sky white.

During the day, the sky changes color.

On a cloudless day, the air gets drier and clearer as the sun rises.

The sky becomes a bright blue.

The sunlight reflecting off the molecules in the air gives the sky its different colors.

At sunset, the sun's light travels through the dusty, lower levels of the atmosphere. The sky shines red.

The wind is a great, flowing current of air.

Force 0: total calm

Force 3: gentle breeze

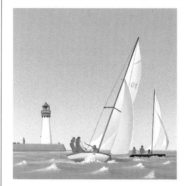
Force 6: strong breeze

Force 8: gale winds

The air around us is always moving. This movement is called turbulence.

The air creates wind as it moves. Why is the air always moving? Warm air rises because it weighs less than cold air. This causes movement and wind. At the seaside, for example, the cold air over the sea moves down to replace the warm air rising over the beach; thus, a sea breeze blows onto the land. At ground level the air may be calm. When smoke rises straight up into the sky, there is no wind. But higher up, the clouds move because of the wind. At very high altitudes, about 7.5 miles (12 km) up, strong air currents called the jet stream encircle the earth. Aircraft try to fly within the jet stream to speed up their journeys. The jet stream can reach speeds of 200 miles per hour (322 km/h).

Earth's air currents as seen from space

A tree distorted by the wind

Windmills use the power of the wind to turn their sails.

Wind can produce a soothing breeze or severe damage.

Force 10: severe storm

Force 12: hurricane

Measuring the strength of the wind

The force of the wind is measured from 0 to 12. Force 0 corresponds to absolute calm, while force 12 corresponds to a hurricane, when the wind speed is more than 74 miles per hour (119 km/h). This method of measuring the wind is called the Beaufort scale after the British admiral, Sir Francis Beaufort, who devised it.

Thunderstorms usually don't get stronger than force 10. Hurricanes at force 12 normally happen along the coasts. They begin over the ocean as tropical storms and are classified as hurricanes when their winds exceed 74 miles per hour (119 km/h).

In the mountains and the Midwest, cold winter winds may produce violent blizzards. Blowing snow from these blizzards can make travel impossible.

A wind generator can produce electricity when the wind turns its blades.

A tornado is a spiraling column of air.

Tornadoes generally develop during severe thunderstorms. The sky turns green and funnel clouds emerge from the clouds, descending to the ground. When these funnel clouds touch ground, they rip apart everything in their path, tearing roofs from homes and uprooting trees. Their winds can exceed 200 miles per hour (320 km/h). Tornadoes may sound like a roaring train and can be frightening, but they usually last only a few minutes.

A tornado

As the clouds grow, it begins to rain.

An old proverb says: rain before seven, fine before 11. In the temperate climate, weather can change in only a few hours. Early morning rain sometimes leads to a clear and sunny afternoon, and then by evening, dark rain clouds may once again gather on the horizon.

Where do clouds come from? There is water in the air, in rivers and lakes, on snow-covered mountains, and in the seas and oceans. About three-fourths of the Earth's surface is covered with water. The sun's heat transforms water into vapor. This vapor then rises with the warm air and collects into clouds as the air cools. Cold air cannot hold as much water vapor as warm air, so the vapor becomes liquid once again and begins to fall back to the Earth as rain.

The water cycle
Rainwater drains into rivers and flows out to the oceans. There the water evaporates, forms clouds, and falls again as rain. Earth's water is never lost in this unending cycle.

Clouds make landscapes in the sky.

Clouds form as moist air rises and cools. To see how this works, have an adult help you with this experiment. Hold a cold plate over a pot of boiling water. Water vapor will rise above the pot and meet the cold plate, where it will once again become a liquid as it cools. This is called condensation. Clouds are made of water vapor that condenses around microscopic dust particles.

Why do raindrops fall from the clouds? Millions of tons of water float around in the form of clouds in the sky. But not all clouds produce rain. The drops only fall if they grow big enough. Droplets collect together to become big drops, until they become too heavy to float and fall as rain. One raindrop is made from about one million cloud droplets.

Many kinds of clouds produce many different shapes. Some clouds are tall and shaped like enormous cauliflower tops; these are the white, fluffy cumulus clouds. Generally, they are associated with fine weather. Sometimes, though, they grow large and dark and form anvil-shaped tops. When this happens, there will probably be a thunderstorm. These cumulonimbus clouds may produce heavy rain, hail, and strong winds.

Flat, horizontal clouds forming thin layers are called stratus clouds. Cirrus clouds are made of ice crystals and take the shape of snowflakes, thin and feathery, or they may look drawn out like threads. At about 3.75 miles (6 km) up, they are called cirrostratus. Altocumulus hover at about 3 miles (4.8 km) up and look like blotches, giving the sky a dappled appearance.

Cirrostratus clouds look like a thin mist and sometimes produce colored halos around the sun or moon. Higher up are the cirrocumulus clouds. They may look like waves or have the rippled appearance of fish scales. People sometimes refer to the "mackerel sky" when these clouds appear.

Look up into the sky and watch these wonderful cloudscapes pass overhead.

1. Cirrostratus
2. Altocumulus
3. Cumulonimbus
4. Cumulus

Lake City Jr. Academy Library

Lightning strikes and thunder rumbles during a thunderstorm.

On hot days, air rises fast. Large cumulonimbus clouds develop and the horizon becomes dark. The wind gets stronger; a storm is brewing. Inside the clouds, water droplets swirl around. This movement creates electricity, and bolts of lightning leap out of the clouds, seeking a path to the ground. The sound of thunder rolls across the sky. It is time to take shelter.

The Greek god Zeus throws thunderbolts across the sky.

Zig-zags of electricity shoot from cloud to cloud, from cloud to ground, and sometimes from ground to cloud. This is lightning. Lightning can be dangerous. It takes the fastest way it can find to get to the ground, traveling down anything high—a church steeple, a tree, a pointed rock, or an antenna. It even strikes water. Some buildings are equipped with lightning rods, which are poles that lead the lightning straight into the ground and keep it from harming the buildings.

Lightning also creates a great deal of noise, called thunder. Thunder is an explosion of air that has been heated by lightning.

Even when the thunderstorm is far away, the lightning is immediately visible. Light travels fast—at 186,000 miles (300,000 km) a second. The sound of thunder takes longer to arrive. Sound only travels at about 1,090 feet (330 m) a second. So it may be several seconds after we see the lightning that we hear its thunder. To work out the distance of the storm, count the seconds between the lightning and thunder and multiply the number of seconds by the speed of sound. If you count 10 seconds, the storm is 10,900 feet (3,300 m) away, or about two miles (3.2 km).

During a thunderstorm, never take shelter under a tree. The safest place to be is inside a stable building.

Rainbows are bridges of colored light across the sky.

Sometimes after rain, the sun comes out from behind the clouds. If it is not too high in the sky, a rainbow appears in a half-circle of multicolored light. An incomplete arc is called a sun dog.

The sun's light looks white to us, but in fact it is made up of seven visible colors. When sunlight passes through a drop of water, the rays of light bend slightly. Each of the different colors of the light bends at a slightly different angle.

Occasionally, the moon is bright enough to produce the same rainbow effect. This is called a moonbow. After a storm, millions of drops of water remaining in the sky break up the white light into a rainbow. Red is always at the outside of the curve, followed by orange, yellow, green, blue, indigo, and finally violet at the inside. You can make your own rainbow with a water hose by standing with your back to the sun and spraying a fine mist in front of you.

The best view of a rainbow occurs with the sun at your back. Mornings or evenings are the best times to see a rainbow.

Sometimes when the water drops are large enough, the sun's rays are reflected twice inside each drop. Two rainbows form, one inside the other, with their colors in opposite order.

Snow is water transformed into ice crystals.

Stars in the snow

If the air gets cold quickly, falling below 32 degrees Fahrenheit (0°C), clouds lose heat and the drops of water inside them turn to tiny crystals of ice. As these ice crystals fall through the sky, they stick together and make starry snowflakes. The snowflakes have intricate designs.

These crystals cling together in a geometric way and always have six branches. The shapes they form depend on the temperature and humidity in the air. More moisture means bigger flakes.

A snowflake consists of ice crystals with air between them. Imagine a pillow stuffed with feathers and a lot of air trapped between the feathers. Snowflakes are much the same way. Freshly fallen snow is very light, but later, as the air is forced out, it packs down into hard ice.

The formation of glaciers

To make a snowman, the snow must be pressed into shape. This pushes the air out from between the snowflakes and allows the snow to hold together. If the temperature dips below freezing the next night, the snowman begins to turn into ice. He will last even after the rest of the snow has melted. This is similar to the way that mountain snow turns into glacier ice.

Up above 10,000 feet (3,000 m) in the mountains of the United States and Canada, cold weather lasts most of the year. Here, when the clouds release moisture, it usually falls as snow. It rarely rains. In this cold climate, more snow falls than the summer sun can melt. Year after year the snow grows thicker, until it becomes compacted and turns into ice. Eventually this packed ice begins to flow downhill, becoming a glacier.

Glaciers are enormous rivers of ice. They flow slowly downhill from the tops of mountains toward the valleys, traveling over the rocky surfaces. Because glaciers are solid, the glacial ice cracks when it goes over a steep slope or around bends in the valley. These cracks are called crevasses and are dangerous for mountain climbers; they are sometimes as much as 200 feet (60 m) deep.

Crevasses open and close as the glacier moves. When the glacier travels down a steep slope on the mountain, it makes an icefall. An icefall is similar to a waterfall formed when an ordinary river goes over a steep slope. The icefall is made of jumbled blocks of broken glacial ice called seracs. The ice in the seracs is white when it still contains air bubbles, turning blue when dense and airless. The older ice is usually the most compact.

How is a weather forecast made?

By observing the clouds and wind, we can guess what kind of weather is on its way. Meteorologists, people who study the weather, use many instruments to help them understand changes in the atmosphere.

People whose jobs are affected by the weather, such as farmers and fishermen, have looked to nature for clues to forecast the weather for centuries. They watched for anxious livestock, crows gathering together, and seagulls flying close to houses as indications of bad weather.

A meteorologist's work

Today, information about the weather comes from around the world at all hours of the day and night. First, a computer processes the data. Then meteorologists draw up maps of the world's weather patterns. By reading the maps, they try to predict what the weather is going to do over the next few days.

Where do meteorologists get the information they need? Weather stations on the ground have barometers to measure the weight of the air, called barometric pressure. When the pressure drops, the barometer falls. This means bad weather is on the way. A weather vane shows the direction of the wind. An anemometer measures its speed. Weather stations also contain a thermometer, which records the highest and lowest temperatures each day. In the United States, 12,000 weather watchers supply data to the National Weather Service.

A weather station

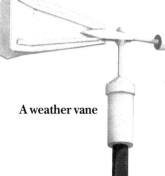
A weather vane

Using satellites and radar for weather forecasting

A hygrometer measures how damp, or humid, the air is. The weather stations also have a rain gauge to collect and measure how much rain has fallen.

Observing the weather from balloons

Special weather balloons are filled with helium and carry instruments high into the atmosphere. The light helium gas takes the balloons to a height of about 19 miles (14 km) above the Earth. The balloons carry a radio transmitter and a radar reflector. Both devices transmit information back to Earth about temperature, atmospheric pressure, and the humidity of the air. Radar trackers on the ground follow the drifting balloon.

Satellites watch the weather, too. They take photos and measure the Earth's weather. Satellites are particularly useful for gathering information about tropical storms and hurricanes. At sea, hundreds of ships voluntarily collect information about the world's weather. This information is used by meteorologists to help them provide accurate weather forecasts.

A satellite

A weather balloon with instruments

Anemometer

Barometer

Why do we need weather forecasts?

We all want to know what the weather will be like. Fishermen need to know how rough the sea will be. Farmers need to know if there will be rain for their crops. Aircraft pilots want to know if the winds will be favorable for flying, and people on vacation want to know if it's going to be sunny.

Our blue planet

Photos taken from satellites show that more than three-quarters of the Earth is covered in water. The Pacific Ocean alone covers nearly one-third of the planet and is larger than all the continents combined. Under the oceans lie volcanoes, mountains, and deep valleys.

Atlantic Ocean

Pacific Ocean

The Pacific, the largest ocean, stretches 9,000 miles (14,500 km) from north to south and 11,000 miles (17,700 km) from east to west.

The Indian Ocean is located between India and Africa. The Arctic Ocean is the most northerly ocean and surrounds the North Pole.

The Atlantic Ocean, shaped like a great letter S, lies between Europe and Africa and North and South America. The Atlantic formed over the last 200 million years as the Americas gradually moved farther and farther away from Europe and Africa. These great continents, called Pangaea, were once connected. A close look at a map shows how their coastlines matched up like pieces in a jigsaw puzzle.

The mysterious depths of the ocean
The ocean floors are far from being flat, boring plains. The Earth's longest mountain chain runs under the oceans, hidden by the deep water. Even before submarines made it possible to explore the depths, scientists knew these mountains were made of volcanic rock. Volcanic activity pushes the continents apart.

Fifty million years separate each of these globes. The third one shows the Earth as it is now. Each year North and South America move about an inch (2 cm) farther from Europe and Africa.

The kingdom of the deep

Since the 1940s, scientists have been exploring the oceans in submarines. They have begun to unlock the secrets of the deep.

More than half the surface of our planet is covered by ocean waters over 6,500 feet (2,000 m) deep. No light penetrates this depth and the water is cold. The temperature is only 35 degrees Fahrenheit (2°C) at 6,500 feet (2,000 m) down. In some places the temperature even reaches 32 degrees Fahrenheit (0°C), but the water doesn't freeze because of its salt content. Without light, no plant life can grow in the deep ocean. The creatures that live there are unlike anything else on Earth. Many species that disappeared from the surface millions of years ago still survive at these depths. Fish in the deep ocean are blind. The lack of light has made their eyesight useless.

Two hundred feet (61 m) down, divers from a mini-submarine film a seaweed-covered wreck.

The ocean waters are always moving.

The Gulf Stream flows northeastward from the Gulf of Mexico.

The waters of the seas and oceans are in constant motion. Currents run through the oceans similar to the currents in rivers. They are pushed along by the winds and flow from places of warm temperatures to places of cold.

The most famous ocean current is the Gulf Stream. Fishermen discovered it long ago while tracking schools of whales. Whales follow the warm water of the Gulf Stream, but never swim in it. The Gulf Stream delivers warm water and a warm climate from the Gulf of Mexico to European shores. Ships use it to help them along on their voyage towards Europe, but when they travel back to America, they go south to avoid it, because it would slow them down.

In 1947, a Norwegian named Thor Heyerdahl used the Humboldt current in the Pacific Ocean to float a balsa wood raft, the Kon-Tiki, from South America to the Polynesian Islands in the Pacific Ocean. It took 101 days.

At the seashore, the sea level rises and falls regularly. This motion is called the tides. The moon and sun's gravities pull the Earth's water towards them, making the sea level rise and fall as the Earth spins. Out in the oceans, the change in water level is not noticeable. At the shore, though, it is clearly visible. The sea level rises for six hours, then, just as slowly, falls again. There are two of these tides every day; the high tide is about 50 minutes later each day. The strongest tides occur twice each month when the sun, moon, and Earth are directly aligned. These are called spring tides.

The Kon-Tiki

Currents, tides, and waves

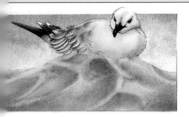 

Sea birds resting on the water bob up and down on the waves. The waves rise and fall, but the water itself does not move forward.

Wind makes waves on the surface of the sea. Ripples on the surface of the water eventually grow to become ocean waves. The waves break against the land at the seashore in a mass of white spray. Out at sea, a big storm can raise waves taller than a house. Ships are built to withstand the violence of such storm waves.

Sometimes waves appear on the sea even without the presence of wind. Waves are born a long way from where they finally come ashore. Waves react in the same way as the ripples that travel outwards when a pebble drops into a pool of water. Waves raised by winds off the coast of the United States may come ashore in Europe, 3,100 miles (5,000 km) away.

Waves only disturb the surface of the sea. Below the surface, fish swim quietly in calm waters. Sometimes, though, undersea earthquakes create waves deep down that ships on the surface know nothing about. Strong winds and powerful spring tides combine to produce tidal waves. These waves don't occur often, but when they do strike they can cause damaging floods along coastal areas.

Tidal surges and tsunamis can be huge, dangerous mountains of water. Tidal surges may be whipped up by violent storms when atmospheric pressure is low. This causes the ocean to rise above its normal level. Tsunamis are caused by earthquakes in or near the sea, or when volcanoes erupt and push aside the sea water. These waves rush over whole islands, crushing trees and houses on their way. In 1896, a tsunami killed 22,000 people in Japan.

Pack ice and icebergs

Most of the Earth's fresh water is in the form of ice at the North and South Poles. Frozen sea water is called pack ice. It builds up during the winter around the seashores near the North and South Poles. In summer, waves break up the pack ice. It floats away, and eventually melts. The ocean at the North Pole is permanently covered with thick pack ice that never melts completely.

A North Pole iceberg

Icebergs look like frozen castles floating on the sea. Icebergs are made of fresh water. They began life as huge glaciers covering the land in the polar regions. Many are born in Antarctica, Greenland, and the mountains of Alaska. When the glaciers reach the sea, they float out over the heavy salt water and eventually break up. These icy fragments, some of which are very large, are called icebergs. The word iceberg actually means "ice mountain."

A South Pole iceberg

North Pole icebergs usually come in strange, pointed shapes; those from the South Pole are often flat.

Most of an iceberg is hidden under water.

Icebergs float around the polar seas and are dangerous to ships passing nearby. Only a small part of the iceberg shows above the sea. More than six times as much of the iceberg is hidden under water. Today, there is an international patrol surveying the danger zones so ships can be warned. One iceberg discovered in 1946 was tracked for 17 years before it melted.

The coasts are battlegrounds between the sea and the land.

Estuary

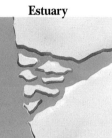
Delta

Along the length of a coastline, the scenery may change every few miles. Some coasts have rugged cliffs plunging down to the sea. Around the next corner, there may be an inlet with a beach. Some coastal areas have flat, sandy, or pebble beaches stretching as far as the eye can see. In the tropics, lagoons sometimes separate offshore islands or coral reefs from the mainland.

Where the river meets the sea

From their inland sources, rivers flow toward the sea. A river may end in a wide mouth called an estuary, where its freshwater pours into the salty ocean.

Lagoon

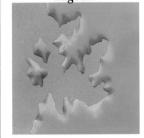
Fjord

Sometimes rivers carry sand and drop it near the sea. Sand banks form and change the paths where the river wanders. This is called a delta. A lagoon is a stretch of calm water protected behind a barrier of land.

The fjords of Norway are deep valleys carved out by glaciers during the last Ice Age. When the glaciers melted, the oceans rose and flooded the valleys.

Coasts are always changing. They may be eroded by the wind, weather, and the sea. They may also be built up by sand and gravel carried by rivers.

Mountains have the power to frighten and fascinate.

There are mountains on every continent. Their lofty peaks, deep, dark valleys, and barren rocks once frightened people. People believed that gods or demons lived on the summits of mountains. The mountains' inaccessibility led people to believe that only the gods could live there. Many of the mountain summits that once frightened people have now been climbed by brave people, and many have died trying to climb mountains.

What is inside the Earth? Like an apple, the Earth has a core. It also has flesh called the mantle, and a skin called the crust. A journey to the center of the Earth would be impossible. We would never be able to survive the tremendous heat and pressure. We live on the thin crust of cool, mostly solid rock that covers the Earth.

Millions of years ago, the continents were joined together as a single, huge land mass. Over time, they have drifted apart. Some still move apart, while others move toward each other.

Below are some of the mountain ranges of the world.

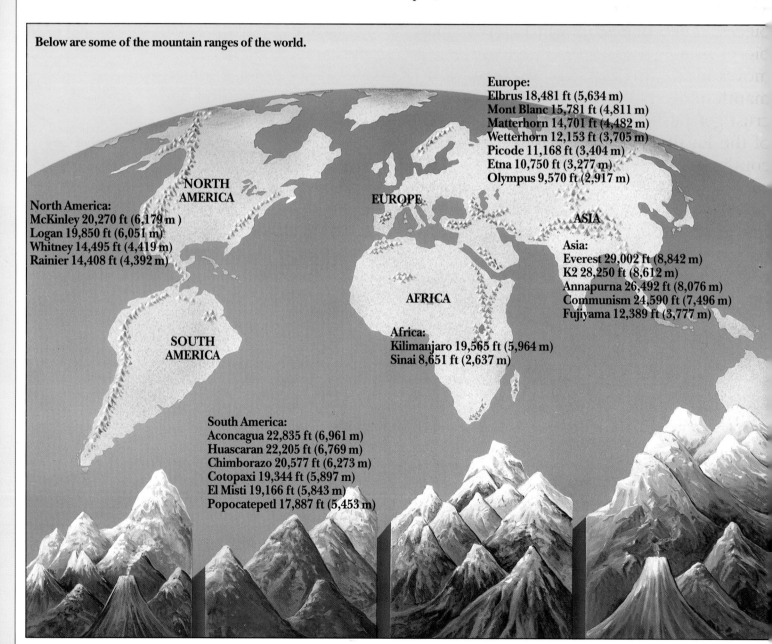

North America:
McKinley 20,270 ft (6,179 m)
Logan 19,850 ft (6,051 m)
Whitney 14,495 ft (4,419 m)
Rainier 14,408 ft (4,392 m)

South America:
Aconcagua 22,835 ft (6,961 m)
Huascaran 22,205 ft (6,769 m)
Chimborazo 20,577 ft (6,273 m)
Cotopaxi 19,344 ft (5,897 m)
El Misti 19,166 ft (5,843 m)
Popocatepetl 17,887 ft (5,453 m)

Europe:
Elbrus 18,481 ft (5,634 m)
Mont Blanc 15,781 ft (4,811 m)
Matterhorn 14,701 ft (4,482 m)
Wetterhorn 12,153 ft (3,705 m)
Picode 11,168 ft (3,404 m)
Etna 10,750 ft (3,277 m)
Olympus 9,570 ft (2,917 m)

Africa:
Kilimanjaro 19,565 ft (5,964 m)
Sinai 8,651 ft (2,637 m)

Asia:
Everest 29,002 ft (8,842 m)
K2 28,250 ft (8,612 m)
Annapurna 26,492 ft (8,076 m)
Communism 24,590 ft (7,496 m)
Fujiyama 12,389 ft (3,777 m)

Earth's mighty internal energy folds and shatters rocks.

How mountains are created
Powerful currents within the Earth's mantle push up on the crust.

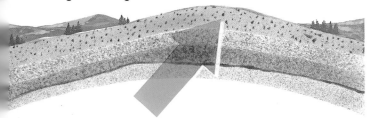

Pushed by the pressure from below, the rocks of Earth's crust fold, crack, or lift up.

The Earth's outer skin, the crust, is not whole. It is made of many different pieces, called plates, that fit together like pieces in a jigsaw puzzle. The mantle moves inside the Earth. Because of these mantle movements, the jigsaw pieces of crust get pushed around over the surface of the Earth. Plates as large as the continents or larger are moving, though very slowly. When two plates push against each other, rocks at the touching edges fold up and crack. The Himalaya Mountains are the result of India moving northwards toward Asia. This collision has been taking place for millions of years.

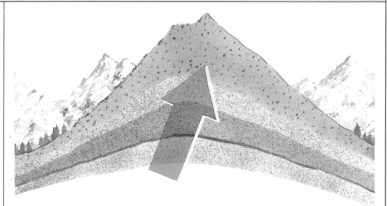

Some mountains grow as much as three feet (1 m) higher every thousand years.

Young mountains erode nearly as fast as they grow. The Himalayas and the European Alps are young mountains. They have developed during the last 40 million years. They are still developing now, but wind, snow, glaciers, and rain wear them away, carrying rocks and stones from the summits into deep valleys. The mountains of Wales and Scotland are old mountains. They stopped developing long ago and their peaks and slopes have become rounded off by erosion over the course of hundreds of millions of years.

Formation of the Himalayas

Mountains form chains stretching across continents.

Life in the mountains of Nepal has made the Sherpa people strong. They can carry loads as heavy as themselves.

Few mountains stand on their own. Usually, they are part of long chains of mountains, one overlapping the next. The Earth has two main mountain chains. The first includes the Alps, the mountains of Greece and Iran, and the Himalaya Mountains. The other chain encircles the Pacific Ocean and includes the Rocky Mountains of North America, the Andes Mountains of South America, the New Zealand Alps, and the mountains of Japan.

Mountain peaks can affect the weather. To get over mountains, clouds must rise higher into the sky. As they rise, they cool. Cold air cannot hold as much water, so it begins to rain or snow. It rains and snows more on the sides of the mountains facing the arriving clouds, while the far sides remain dry. Snow and glacier ice from mountains provide reserves of fresh water for many streams, rivers, and lakes.

The Himalayas and the Andes: mountain fortresses

Lake Titicaca, at nearly 12,500 feet (3,810 m) up in the mountains, has been a center of South American civilization for hundreds of years. Incan ruins have been found there.

Mountain scenery changes with elevation and with the amount of sunshine on the slopes. The weather becomes colder as the elevation increases. Trees are smaller higher up, gradually giving way to shrubs and grasses of the alpine meadows. Higher still, only mosses and lichens can withstand the cold. At the very tops of some mountains only snow covers the land. Here glaciers are born.

In spite of the intense cold, some people live in the highest mountain valleys. On the rainy side of the Himalaya Mountains, people have carved terraces into the slopes to cultivate as much of the land as possible. They use the higher mountain plateaus and slopes as pasture for rearing and grazing ponies, sheep, and yaks. At 29,002 feet (8,842 m), Mount Everest in the Himalayas is the highest mountain in the world. It forms the border between Nepal and Tibet.

The Andes mountains stretch over 4,500 miles (7,200 km) from Venezuela in the north to Tierra del Fuego at the southern tip of South America. The Andes are young mountains and are home to many people despite the frequent earthquakes and volcanoes.

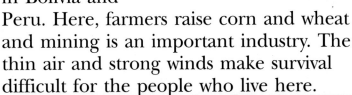

The people of Bolivia and Peru live and herd their llamas as high up as 14,000 feet (4,300 m).

These high mountains surround a vast high plain, the altiplano, in Bolivia and Peru. Here, farmers raise corn and wheat and mining is an important industry. The thin air and strong winds make survival difficult for the people who live here.

Machu Picchu in Peru, built by the ancient Incas as a fortress city about 500 years ago, was only rediscovered in 1911.

Volcanoes are mountains that breathe fire.

Volcanoes shudder and throb with escaping hot gases, ash, and molten rock. Rocks from the crust were forced down into the mantel by the movement of the plates. These rocks deep inside the Earth have melted. Sometimes pressure builds up in the mantle and forces the molten rocks up through cracks in the crust. When this happens, a volcanic eruption occurs.

Inside a volcano: 1. Reservoir of magma 2. Chimney 3. Cone with crater at top

Temperatures within the mantle are hot enough to melt rock. When this rock melts, pockets of molten rock called magma form. The magma forces its way up through cracks in the Earth's surface. The magma spurts out into the air, pushed by the pressure and hot gases from inside the Earth. A volcanic eruption is happening.

Lava is red when it is hot. As it cools, it turns gray or black.

When magma escapes from the Earth, we call it lava. If lava is thick and sticky, it becomes solid at the crater, or is hurled up into the air in the form of boulders, ash, and fine dust. If the lava is liquid and runny, it flows like a river down the sides of the cone, or makes a red hot lake in the crater.

The life cycle of volcanoes

In 1943, near a small village west of Mexico City, a farmer working in his field felt the ground shake.

Rarely does someone see a volcano when it first appears. Only one or two volcanoes are created each century.

The next day, the newly created Paracutin volcano stood where the field had been. Within a week it had grown 460 feet (140 m).

In 1963, Surtsey, a new volcano, burst out of the sea off the coast of Iceland, creating a new island. Surtsey erupted for four years, with hot lava and ash pouring into the sea. Predicting where the next new volcano will occur is difficult.

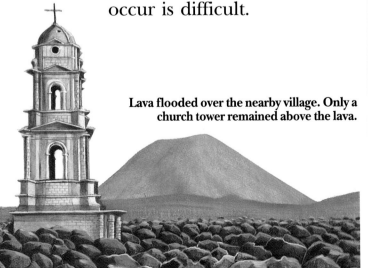

Lava flooded over the nearby village. Only a church tower remained above the lava.

A volcano is extinct, or dead, when it hasn't erupted for tens of thousands of years. Sometimes, though, a volcano that seems extinct may simply be dormant, or asleep. One day it may wake up and erupt again. Between eruptions, wind, snow, and rain wear away the cone. By measuring how much has worn away, scientists can tell how long it has been since the volcano last erupted.

The reawakening of a dormant volcano can be spectacular. In 1980, a volcanic explosion blew off the top of Mount St. Helens in the western United States. This one explosion reduced the height of the mountain by 1,500 feet (460 m).

Cotopaxi, a volcano in Ecuador, rises 20,556 feet (6,267 m) above sea level, making it the highest land volcano on Earth.

A dormant volcano may be cool enough for water to form a lake in the crater. Sometimes the water in these lakes is warm or even hot. Some old volcanoes have large craters many miles across. Especially large craters form when the top of a mountain collapses as the magma empties out of it. Sometimes the force of the eruption literally blows the top off the mountain. These large collapsed craters are known as calderas.

Lake City Jr. Academy
Library

In search of the secrets of volcanoes

More volcanoes exist in the oceans than on land. Much of the sea bed is made up of lava formed in the last 200 million years. These are among the youngest rocks on Earth. Cooled quickly by the sea water, lava sometimes hardens into strange shapes. Called pillow lava, these rounded masses are about three feet (1 m) in diameter. Deep in the ocean, volcanic gases rich in metals like zinc and lead also escape from cracks in the sea bed.

Fighting the power of a volcano

As a volcano threatens to erupt, people and animals can usually do nothing more than get out of the way. When a volcano in southern Iceland awoke after 5,000 years of dormancy, the people of this island managed to escape danger, although their homes were buried under ash and lava. To stop the lava flow that threatened to block a town harbor, the Icelanders doused it with water from high pressure hoses. This hardened the lava and stopped it in its tracks.

If we want to reduce the danger of volcanic eruptions, it is important to learn when the volcanoes are likely to erupt. This is the task of volcanologists.

Volcanologists are the doctors of volcanoes. They listen to volcanoes in an attempt to judge when the next eruption will happen. Like a doctor using a stethoscope and a thermometer to understand the health of a patient, volcanologists use various instruments to study the movement of magma inside the volcano. They measure any swelling of the slopes of the volcano's cone and the temperature of the volcano's gases to discover whether the volcano will become active.

Hardened lava from earlier eruptions can be bulldozed to make a wall to divert lava flows.

Clouds of gas, rich in metals, pour from undersea cracks.

There are thousands of active volcanoes on Earth.

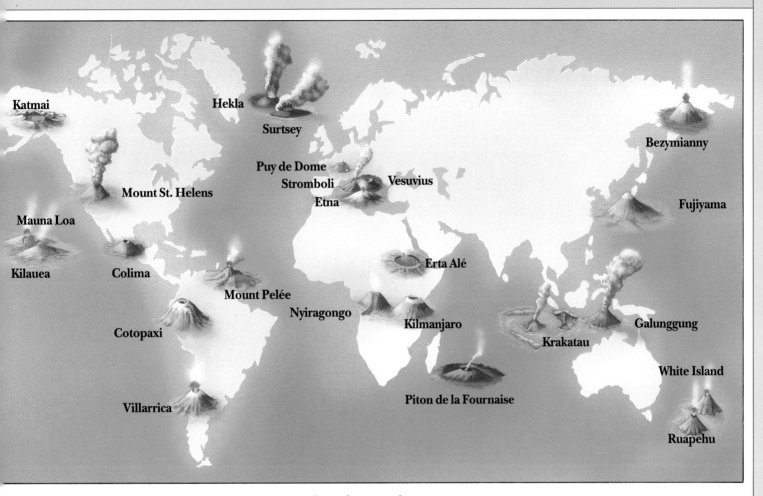

Some famous volcanoes

One of the deadliest eruptions this century took place in Martinique in 1902. A huge cloud of ash, rocks, and hot gases from the eruption of Mount Pelée engulfed the town of St. Pierre, killing roughly 30,000 people within a few minutes.

Earthquakes happen when the Earth's crust moves.

Earthquakes happen all over the world in areas called seismic zones. Seismic zones occur where the plates of crust covering the Earth's surface meet each other. Inside the Earth, the mantle is always moving, which in turn moves the plates. These plates push against each other, building up tension between them. When the tension between plates becomes too great, they grind against each other, causing the Earth's surface to tremble and shake.

Earthquakes occur constantly on Earth. Most are small and cause little damage. Few earthquakes every year are severe, killing people and destroying homes.

As the plates grind against each other, rocks break and shock waves travel outwards, making the ground tremble.

An earthquake

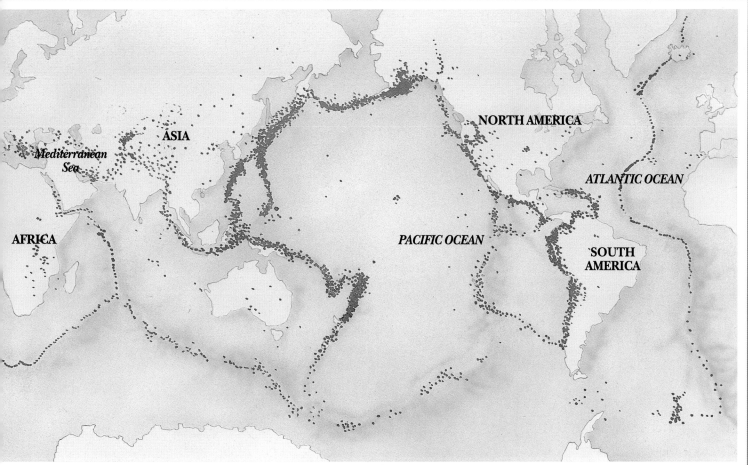

The most active seismic zones form a ring around the Pacific Ocean. A second zone runs from the Mediterranean to China. Others run along the mountain chains of the deep oceans.

Where do earthquakes begin? Earthquakes start from a central focus, sometimes as far as 435 miles (700 km) beneath the Earth's surface. Shock waves travel out in all directions from the focus, including upward to the surface. The place at the surface immediately above the focus is called the epicenter.

To record earthquakes, scientists use instruments called seismographs. Seismographs make a record of the size and shape of the shock waves traveling out from the earthquake through the crust. From the information the seismograph records, experts can classify earthquakes using the Richter scale. This scale assigns an earthquake a number based on its strength. An earthquake classified as a one would be weak, while an eight would be strong. In 1906, an earthquake measuring 8.3 on the Richter scale devastated the city of San Francisco.

People living in countries where earthquakes are common know it's safest to get under a heavy desk or table when the shaking begins.

A hidden world under the ground

In limestone country, spectacular caves form underground. Some caves grow enormous, larger than the biggest cathedrals and office blocks. Rivers flow through these underground caverns, cascading in sparkling waterfalls to form pools and vast lakes. A whole world exists beneath our feet.

Water plays a large part in creating caves. When it rains, much of the water seeps into the ground. In areas where limestone is present, the acid in rainwater eats away the rock over millions of years. The limestone dissolves along the paths made by the water, leaving caves. The water eventually reaches a place where there's no more limestone. The water then becomes an underground river, flowing downhill, just like a river above ground. Eventually, the water comes out into the light of day as clear, clean spring water, perhaps at the bottom of a deep valley. Volcanic activity may also create caves when gas pockets form in the lava or ice beneath the lava melts.

Strange shapes grow in caves. Water drips off the cave roof and splashes on to the floor below. Slowly, over thousands of years, the minerals in the water build up to form beautiful icicle-shaped structures. These structures hang from the roof or grow up from the floor where the drips fall. In time, the stalactites growing downwards meet the stalagmites rising from the floor of the cave.

Speleologists explore and study underground caves.

Some caves look like a fairyland with shapes taking the form of organ pipes, melting candles, tree trunks, and draped curtains. The great columns that seem to hold up the roofs of caves amaze most visitors. This underground landscape was created by dripping water.

Speleologists study a whole underground world—the caves themselves (geology), the underground rivers (hydrogeology), and the animals and plants in the caves (zoology and botany). Speleologists, for their own safety, need also to know how humans react to cold and damp in caves, because they spend a lot of time in them.

Exploring underground

Speleologists need to prepare for the many dangers that exist in dark, unexplored caves. Cave explorers carry sophisticated equipment to help them through these dangers: ropes, rope ladders, lights, inflatable rubber boats for following underground rivers, and radios to stay in contact with each other and with their support team above ground.

The endless work of water, wind, and frost...

Rock towers in Utah

If you build a sand castle on the beach, it eventually gets washed away by the waves. If you dam a river with some pebbles to make a little pool, these too will be swept away by the river one day. These forces and others continuously sweep, reshape, and erode the surface of the Earth.

When cliffs are worn away by waves, parts of them collapse. The loose rocks become pebbles, which the waves toss around, wearing the cliffs away more.

Along the coast, waves beat against the shore. Waves have enormous power. They lift pebbles and boulders, hurling them against cliffs. In time, erosion wears away even the hardest rock. Where the rock is soft, this constant wear forms caves, bays, and inlets. Harder rock takes longer to erode and might take the shape of steep rugged headlands, standing out against the battering of the waves.

Rainstorms reshape the land. In a storm, rainwater percolates through the soil or runs away over the rocks. In the mountains it pours as torrents over the rocky ground. Wherever it goes, water carries with it some rock, mud, or sand grains. A heavy flood can move whole mountainsides. These landslides carry away broken fragments of rock and anything else in their path.

Underground water in Borneo has worn away a great mountain of limestone: all that remains are these strange pinnacles standing above the forest.

...constantly reshapes the Earth's surface.

In the mountains, ice plays a large part in wearing away the rocks. Snow and rainwater seep into cracks in the rock, turning to ice in the cold, mountain weather. As water freezes, it expands with a force strong enough to break rocks apart. Little by little, over millions of years, mountains are worn away by this process. Eventually they are no longer recognizable as mountains.

In the deserts where there is little water, wind wears away the rocks. Winds pick up grains of sand. Even in a breeze, the wind acts like a sandblaster on rocks, constantly eroding them. The sand dunes that form because of this erosion may be long ridges running parallel to the direction of the wind, or crescent shapes with their points facing into the wind.

The wind and rain carry all this eroded rock and soil down to the valleys and plains. In time, even the deepest and widest valleys will be filled by this leveling process. At the foot of every mountain stream rests fan-shaped deposits of broken rock. Glaciers leave behind a moraine, or accumulation, of rocky fragments. In the desert, the wind creates sand dunes. In time, these rock fragments will make new rocks and the process will begin again.

All across the Earth's surface, rock is being worn away and new rock is being made. Every day, erosion exposes rock formations millions of years old. At the same time, new rock is thrust up from beneath the Earth's surface. The surface of the Earth is truly in perpetual motion, always recycling itself.

These old volcanoes have been extinct for a long time. Wind and rain have rounded off their summits and made their slopes more gentle.

Rainwater even wears away granite, one of the hardest rocks. It seeps between the tiny grains of the rock, freezes, and forces them apart.

A map of the Earth today

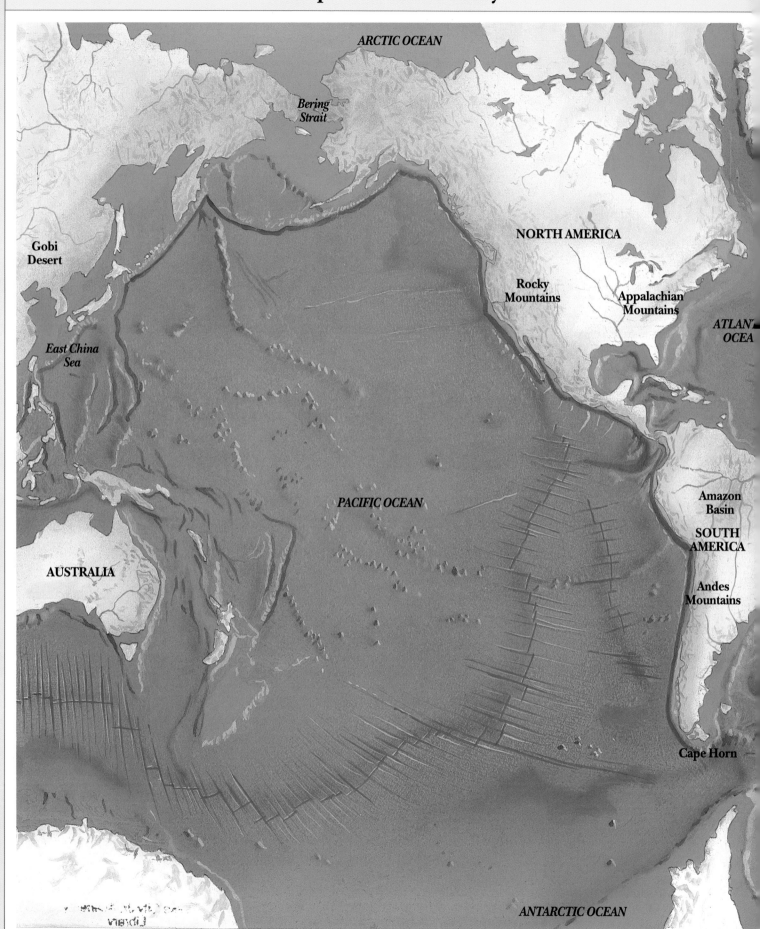

58

The Earth never stops changing.

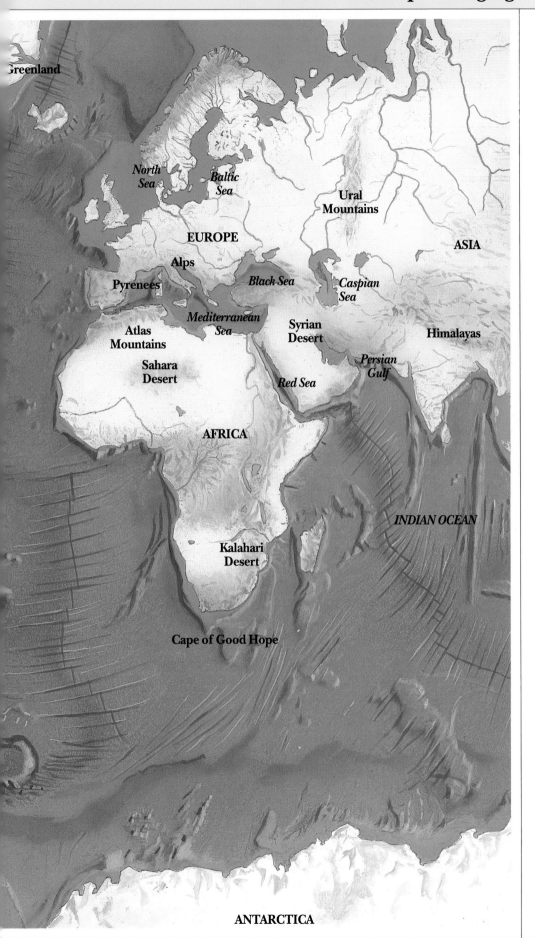

Earth is a living planet that is always changing. Because most of the changes happen so slowly, we do not realize they take place. Earthquakes and volcanic eruptions remind us in a brutal way that the Earth is alive. We know, even without seeing it, that mountains grow at the same time they are being worn away. New rocks are thrust up from deep within the Earth to replace those destroyed by the forces of erosion. Whole continents move, some getting closer together, others moving apart. Climates change, too. Less than a hundred thousand years ago, much of Europe and North America was covered with glaciers. Human civilizations have only existed on the Earth for a few thousand years, which is only a brief instant in the history of the Earth.

Lake City Jr. Academy
Library

EXPLORE AND PLAY

Games and activities, intriguing facts, a quiz, sayings, a glossary, and addresses of places to visit, followed by the index

Did you know?

The distance from the Earth to the sun is 93 million miles (148 million km).

The diameter of the sun is 110 times greater than the diameter of the Earth. The sun's volume is 1,300,000 times greater than the Earth's.

How does the moon's shape determine whether it's on the way to being a full moon or a new moon? When the moon makes a "D" shape, it is growing toward full.

Winter in the Northern Hemisphere is the shortest season, lasting 89 days. Since the Northern and Southern Hemispheres have opposite seasons, summer is the shortest season in the Southern Hemisphere.

Not only are mountains high, they are heavy as well. Their bulk adds so much extra weight to the Earth's crust that the entire thickness of the plate sinks down a little, into the soft mantle below. The mountains would be even higher if they did not sink in.

Why does the sun look red when it rises and sets? A layer of dust always hangs above the surface of the ground. In the mornings and evenings, the rays of the sun reach us at an angle, traveling a long distance through this dust. The dust scatters the blue light away, and only the red reaches us.

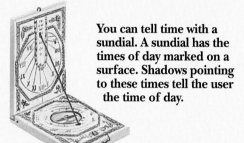

You can tell time with a sundial. A sundial has the times of day marked on a surface. Shadows pointing to these times tell the user the time of day.

People learned a long time ago how to tell the time of day from the sun by using a stick planted in the ground, called a gnomon. For hundreds of years the gnomon was the only way of telling the time. The sundial was just an improvement of the gnomon.

You could use your shadow to tell the time, but you'd need to stand still for a long time. You would know when it was midday, because that's the time when your shadow would be shortest—when the sun reached its highest point in the sky.

Mark where your shadow falls at various times of day, and you can become a sundial.

Liquid water and ice cover about 75 percent of the Earth's surface. The average depth of the oceans is 12,500 feet (3,800 m). If the Earth's surface was completely level, water would be 2,500 feet (760 m) deep everywhere.

The North Pole is permanently covered in a thick sheet of ice. In 1909, Frederick A. Cook and Robert E. Perry both claimed to be the first to reach it. But in 1958, the nuclear submarine, *Nautilus*, became the first vessel to travel under the North Pole.

Thunderstorms
In 1752, Benjamin Franklin proved that lightning wasn't a great fire in the sky as some people had thought. By flying a kite in a thunderstorm, he proved that lightning was really electricity. This was a dangerous experiment, but it later led him to develop lightning rods to protect buildings in storms. His invention has saved many homes from fires started by lightning.

Legends about the universe

Throughout time, legends have existed to explain how the Earth came to be. The people who made up these legends looked at things differently from the way we do now. The ancient Scandinavians believed in Thor, the god of thunder. They thought he hit the sky with his hammer when he was angry, making thunder and lightning.

Thousands of years ago, the creation of the Earth was described in stories about the lives of various gods. The Greeks said that once upon a time there was nothing, and out of this came Gaia, the Earth, and then Uranus, the heavens, the first ruler of the universe. Their twelve children were the Titans. Cronos, their youngest son, wounded his father, so he could rule the universe instead. To make sure nobody got rid of him, Cronos tried to eat his own sons.

There are countless legends about the sun:

The sun god of the ancient Sumerians came out of a cave each morning and rode across the sky in a brilliant chariot of light. Each evening he drove the chariot back into the mountain.

For the ancient Egyptians, the sun god was the hawk-headed Ra. Another sun god was the falcon-headed Horus, who took earthly shape in the pharaohs. The Egyptians also had different names for the sun as it appeared at different times; Khepri was the rising sun, Atum was the setting sun, and Re was the sun at its height.

Many mountain legends also exist:

People held mountains in awe because they believed the gods lived there. In many countries, people still travel long distances, through great dangers and intense cold, to visit their gods in the mountains.

In Peru, millions of indigenous people go on a pilgrimage each spring into the Andes Mountains. In the bitter cold, they make their way to a sanctuary near the town of Cuzco. The Ukuku tribes take home some glacier ice, which they believe cures various illnesses.

Cronos' wife, Rhea, wanted to save her youngest son, Zeus, from being killed by the ambitious Cronos. She tricked Cronos into swallowing a stone instead of eating Zeus, who grew up hiding in a cave. Later, he forced his father to spit out all the children he'd eaten. Cronos and the Titans were sent to Tartarus, the lowest region of the underworld, and Zeus became supreme ruler of the world.

The people of ancient times believed Zeus and the other gods and goddesses lived at the summit of Mount Olympus in Greece. From there, Zeus threw thunderbolts at anyone who angered him. Other gods and goddesses who lived on Mount Olympus were Demeter, the goddess of the harvest; Apollo, the god of music and prophecy and the most handsome of all the gods; Aphrodite, the goddess of love; and Artemis, the goddess of the hunt. Poseidon, Zeus's brother, ruled over the sea.

■ **Records**

The largest meteorite shower ever recorded occurred in Arizona in 1966. Nearly 2,300 meteorites passed overhead in just 20 minutes, streaking the sky.

Earth's nearest neighbor in space is the moon. In 1609, Galileo became the first person to look at the moon, using a telescope he invented.

The first man-made device to reach the moon was a Russian spacecraft in 1959 named Luna II.

The highest elevation reached by humans on the moon was 25,688 feet (7,831 m) in the Descartes Highlands. This record was achieved in 1972 by astronauts John Young and Charles Duke.

In 1955, the longest solar eclipse of the sun ever measured happened in the Philippines and lasted seven minutes and eight seconds.

A jet traveling at the same speed as an eclipse can watch it longer. A Concorde jet followed an eclipse of the sun for 72 minutes in 1973.

In 1519, Ferdinand Magellan set sail on an expedition that circled the world for the first time. It took three years. Out of the five ships that set out to sail around the world, only one returned to Spain. This journey proved that the Earth was round. Later, in 1687, Newton proved that the Earth is slightly flattened at the two poles.

Jupiter is the largest planet in the solar system, with a volume 1,323 times that of Earth.

Mercury makes the quickest journey around the sun— taking only 88 days.

The Earth's highest mountain is called "Chomolungma" by the Tibetan people. The name means "goddess mother of the world." We know it as Mount Everest, after George Everest, the first person to measure its height. In 1987, another mountain in the Himalayan range called K2 was remeasured. K2 briefly became the highest mountain. The people who had climbed it suddenly became the people who had climbed the highest peak on Earth. But the new measurement proved to be false. Mount Everest remains the highest mountain in the world. Its summit was first reached in 1953 by Sir Edmund Hillary of New Zealand and his Sherpa guide, Tenzing Norgay of Nepal.

The highest mountain in Europe is Mount Elbrus in the Caucasus Mountains, standing at 18,480 feet (5,630 m) high.

The highest point in North America is Mount McKinley at 20,320 feet (6,197 m). Mount McKinley is in Alaska's Denali National Park.

The largest desert in the world is the Sahara in North Africa. This desert is 20 times larger than Great Britain.

The largest active volcano is Mauna Loa in Hawaii. Since 1832, it has typically erupted an average of every three and a half years.

The world's largest glacier is the Lambert Glacier in Antarctica, measuring 250 miles (400 km) in length and 40 miles (64 km) in width.

The fastest moving glacier is the Quarayac in Greenland, traveling 60 to 80 feet (18 to 24 m) a day.

The largest iceberg ever observed covered 12,000 square miles (19,200 sq km), an area larger than the state of Massachusetts.

The world's largest lake is the Caspian Sea, on the border between Europe and Asia.

The highest clouds are the cirrus, which develop at about 20,000 feet (6,000 m) above Earth's surface. The lowest clouds are the stratus at about 6,500 feet (2,000 m) and lower.

The biggest clouds are the cumulonimbus, which produce violent thunderstorms and can grow several miles high.

The largest island is Greenland. The next largest is New Guinea.

The largest rivers
By volume of water, the largest river is the Amazon. It is 4,007 miles (6,448 km) long and flows from Peru, across Brazil, to the Atlantic, where its estuary is 50 miles (80 km) wide. It has hundreds of tributary rivers.

By length, the longest river is the Nile, which flows into the Mediterranean. At 4,145 miles (6,670 km) long, it is the only river longer than the Amazon.

The longest river in the United States is the Mississippi at 2,350 miles (3,781 km) in length. It begins in the lake country of northern Minnesota and empties into the Gulf of Mexico.

The deepest part of the ocean is the Marianas Trench in the Pacific Ocean. The deepest part of this trench is 35,837 feet (10,924 m) below sea level.

The time zones

When it is noon in Denver, it is nighttime in India and people are asleep, and in New Zealand it is breakfast time. Why isn't it the same time everywhere?

Sunrise and sunset vary according to people's locations. All over the world, people set their clocks to agree with the timing of day and night for their own area. In 1883, railroads in the United States began using time zones to make sure their trains would be on time.

Calculating time zones around the world

Time for around the world is based on the zero meridian at Greenwich, England. At the zero meridian, the world is divided into 24 zones, one for each hour of the day. Each time we go from one zone to the next, the time changes by one hour. For example, between New York and London there are five time zones, resulting in a difference of five hours.

Because the Earth spins in an easterly direction, the sun rises earlier the further east you go. As you go east, you need to add hours to the time. When you go west, you subtract hours. When it is noon in London, it is nine in the evening in Japan, seven in the morning in New York, and four in the morning in California.

Sometimes, for convenience, a country may decide to alter its official time from what it should be for its time zone. They may do this so they can have the same time as the countries next door, making telephone communication easier. France and Spain have decided to have the same time as the rest of continental Europe further east. Only Great Britain and Portugal have a different time than the rest of Europe.

Using this map, you can figure out what time it is in any country in the world.

To make the time zones,

the Earth is divided into slices shaped like the segments of an orange. The imaginary lines that separate the segments are called meridians.

Each segment, or piece, is a time zone in which everyone should have the same time. There are 24 time zones, one for each of the 24 hours in a day. This is how long the Earth takes to make one complete rotation. When crossing the International Date Line, a day is subtracted when traveling east. One day is added when traveling west across the International Date Line.

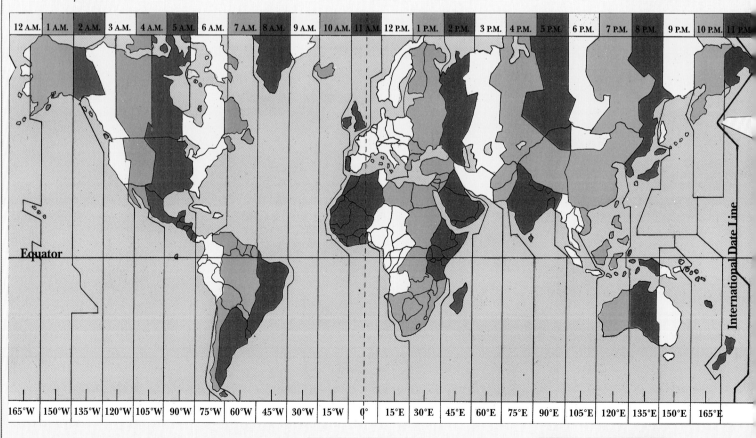

What will the weather be like tomorrow?

People who work the land, live in the mountains, or sail the seas look for signs to help them predict the weather.

Pine cones are very sensitive to dampness in the air. The scales on the cone close up when it is going to rain and open in dry weather.

Some people look for rain when their chairs start to squeak. Others believe rain will come when leaves show their undersides.

People sometimes watch animals for signs of threatening weather:

When bad weather approaches, cows lick themselves and horses paw the ground. Bees return to their hives.

Some animals are good weather forecasters.

If it's going to be wet and stormy, spiders reinforce their webs, or make them smaller, and hide themselves away. When fair weather is near, they spin bigger webs. It will be a clear night if they do this in the evening. If a wooly bear caterpillar's center band is wide and dark, a mild winter is on the way.

If a frog sits out on a lawn or field, it means the air is moist, and it may soon rain. If it hides away under a stone, the air is dry, and sunshine is on the way. Birds flying low often indicates a change in weather.

Donkeys bray continuously when foul weather approaches.

Flies, mosquitoes and horse-flies bite much more in wet weather.

A west wind means good weather, while an east wind means rain may soon be on the way.

Worms and snails come out into the garden during wet weather.

Some people say that cats wash behind their ears when the air is moist.

When bad weather approaches, moles get busy making their hills higher for protection.

Hens scratch the dust and collect their chicks together when rain is near.

The giants of astronomy

These are just a few of the people who made great discoveries about the skies.

Ptolemy lived in Alexandria in ancient Egypt. He believed the Earth was the center of the universe. For over a thousand years, no one questioned his views on astronomy.

Nicolaus Copernicus (1473–1543) finally disproved Ptolemy's ideas. He proved that the Earth spins in space and travels around the sun once a year, just like the other planets.

Johannes Kepler (1571–1630) proved that the planets don't travel in circles, but make ellipses around the sun. He explained this by using mathematical laws.

Quiz

Each question has only one right answer. The answers are at the bottom of the next page.

1. Which planet is closest to the sun?
a. Mars
b. Mercury
c. Earth

2. The moon is
a. bigger than the Earth.
b. the same size as the Earth.
c. smaller than the Earth.

3. The equinox is the time when
a. the days are the longest.
b. the days are the shortest.
c. day and night are of equal length.

4. In summer, the Earth goes around the sun
a. more slowly.
b. faster.
c. at the same speed as the rest of the year.

5. The time just before night is called
a. dawn.
b. dusk.
c. the aurora.

6. When it's nine in the morning in London, what time is it in New York?
a. four in the morning
b. eight in the evening
c. three in the afternoon

7. You hear thunder
a. at the same time as you see lightning.
b. after you see the lightning.
c. just before you see the lightning.

8. The tides are caused by
a. the rotation of the Earth alone.
b. the pull of gravity from the moon and the sun.
c. the rotation of the Earth around the sun.

9. An iceberg is
a. a piece of pack ice.
b. ice broken off a glacier.
c. a block of rock covered with ice.

10. The time span from one full moon to the next takes
a. 31 days.
b. 27 days.
c. 30 days.

Galileo Galilei (1564–1642) was the first person to look at the sky through a telescope. He saw things that no one had ever seen before, such as the moons of Jupiter and the rings of Saturn.

Isaac Newton (1642–1727) was physicist, mathematician, and astronomer. His law of gravitation explains why an apple falls to the ground, and why the moon goes on circling the Earth.

Albert Einstein (1879–1955) was one of the greatest scientists of the 20th century. His theories changed the way people think about space and time. Eventually, his ideas also led other scientists to develop atomic energy.

People have long dreamed of exploring the depths of the oceans.

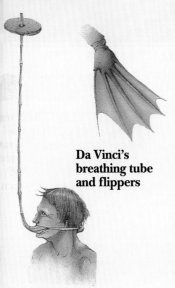

Da Vinci's breathing tube and flippers

In the 15th century, Leonardo da Vinci, an artist, architect, and engineer, designed flippers and a mask with a breathing tube.

11. Which planet is farthest from the sun?
a. Venus
b. Pluto
c. Earth

12. How many years does it take the Milky Way to make one complete rotation?
a. 220 million years
b. 1 million years
c. 5 billion years

13. We measure the weight of the air with
a. a barometer.
b. a thermometer.
c. an anemometer.

14. Which is the outside color on a rainbow?
a. red
b. yellow
c. indigo

15. A hurricane has winds measured at
a. force 9.
b. force 10.
c. force 12.

16. Which ocean is the largest?
a. the Atlantic
b. the Pacific
c. the Antarctic

17. The sun rises
a. in the east.
b. in the south.
c. in the west.

18. The nearest planets to Earth are
a. Venus and Mars.
b. Mercury and Mars.
c. Jupiter and Saturn.

19. A shooting star is
a. a star traveling across the sky.
b. the shining trail of a meteor.
c. dust particles from the sun.

20. Atmospheric pressure is
a. greatest near the ground.
b. greatest on tall mountains.
c. the same at all heights.

Answers: 1 b, 2 c, 3 c, 4 a, 5 b, 6 a, 7 b, 8 b, 9 b, 10 b, 11 b, 12 a, 13 a, 14 a, 15 c, 16 b, 17 a, 18 a, 19 b, 20 a

The first diving helmet (1837)

In the 19th century, a French writer named Jules Verne wrote a story about Captain Nemo, a man who lived 20,000 leagues under the sea. At such enormous depths, Nemo's submarine was a fabulous palace for watching the fantastic creatures of the deep.

In reality, submarine exploration began with diving bells, which allowed people to go down to 200 feet (60 m). Bathyscaphs, water-tight cabins which could be navigated, allowed much greater depths. The first, invented by Frenchman Auguste Piccard, went as deep as 10,330 feet (3,150 m) in 1953. In 1960, Picard's son, Jacques, reached a depth of 35,800 feet (10,920 m) in the Marianas Trench.

A diving machine (1776)

Lake City Jr. Academy Library

■ **Glossary**

Air: the gases and water vapor that make up our atmosphere. The exact proportions of the gases vary according to temperature, altitude, and pollution.
Antarctica: the ice-covered continent situated mainly within the Antarctic Circle and surrounding the South Pole.

Arctic: the region between the North Pole and the timberlines of North America, Europe, and Asia.
Astrologer: one who studies the positions of heavenly bodies in the belief that they influence events on Earth.
Astronomy: the study of the movement and nature of planets, stars, and galaxies.

Atmosphere: the gases that surround and protect the Earth.

Bathyscaph: a vessel used for undersea research.

Bay: a body of water partially enclosed by land and providing access to the sea.

Carbon dioxide: a gas that is a small but important part of our atmosphere. Animals breathe it out, and plants absorb it in daylight. Carbon dioxide also traps heat from the sun near the Earth's surface.
Climate: the usual patterns of weather for a region.
Comet: a frozen ball of gas traveling in an orbit around the sun. Comets have a long, shining tail produced by the heat of the sun.
Condensation: the process of water vapor turning to liquid. This usually occurs when the vapor comes in contact with something cold.
Continent: one of Earth's main land masses. There are seven continents: Africa, Antarctica, Asia, Australia, Europe, North America, and South America.
Core: the dense metal at the center of the Earth, consisting mainly of iron. Much of the core is molten, but the center is thought to be solid.
Crust: the outer solid layer of Earth, about 3 miles (5 km) thick in the oceans and 22 miles (35 km) thick on the continents.

Desert: a barren, sometimes sandy, region of little rainfall where plants and animals are rare.

Eclipse: an event that occurs when the light from a moon, planet, or other object in the sky is blocked by another one.
Epicenter: the point on the Earth's surface above the focus, or place where an earthquake begins far under the Earth's surface.
Equator: an imaginary line equally distant from the poles that divides the Earth in half.
Equinox: the time of year when day and night are of the same length.
Erosion: the gradual wearing down of rocks and soil by wind, water, or other causes.
Evaporation: the process of a liquid turning into a gas. This usually occurs when the liquid has been heated.

Galaxy: a group of stars, gas, and dust in the universe.

Gravity: a force that pulls objects toward Earth.
Greenwich meridian: an imaginary line passing through Greenwich, London, upon which all the time zones are based.

Hemisphere: the two halves of the planet on either side of the equator. Earth has a northern and a southern hemisphere. Also, half of any sphere.

Hydrogen: a highly flammable, lightweight gas. It is the lightest of all gases and the most abundant element in the entire universe.

Ice age: a time when our climate was extremely colder and much of the Northern Hemisphere was covered with ice. The last ice age ended about 10,000 years ago.

Jet stream: strong air currents traveling in a generally westerly direction above the Earth at speeds of up to 250 miles per hour (400 km/h).

Legend: a story that passes from generation to generation, often by storytelling rather than by writing.
Lunar: a word used in reference to the moon.

Mantle: the rocky part of the Earth beneath the crust and surrounding the iron core.
Meteor: the name given to pieces of rock and metal traveling through Earth's atmosphere.
Meteorite: a piece of rock or metal that has fallen to the Earth from space.
Milky Way: the name of our galaxy. It can also be called the Galaxy, with a capital "G."
Moisture: a small amount of water turned into gas and mixed into the air.
Month: a calendar month is 30 or 31 days (except for February); a lunar month—the circling of the Earth by the moon—is about 28 days.
Moon: a natural satellite in orbit around a planet.
Mouth: an opening where a river empties into a larger body of water.

Ocean: one of the large bodies of salt water that cover 70 percent of the Earth's surface.

Orbit: the journey a moon, planet, star, or galaxy makes around another object in the universe. The Earth's orbit around the sun takes one year.
Oxygen: a gas making up about one-fifth of the Earth's atmosphere. All forms of animal life need oxygen to live: we breathe it in the air; fish breathe it in the water.

Pack ice: Sea water that freezes. During winter, pack ice builds up around the poles, and in summer it breaks up and floats away.
Phases of the moon: the shapes of the moon that we see repeated, such as the crescent, the half, or the full moon.
Plates: pieces of the Earth's crust and mantle that fit together and cover the entire surface of the Earth.
Poles: the most northerly and southerly parts of the Earth. The Earth spins on its axis, the imaginary line that runs through the poles.

Radar: an electronic instrument using radio waves to locate any moving or fixed object.
Richter scale: a device used to measure the intensity of an earthquake by recording ground motion.

Satellite: a small object in the universe that orbits around another, larger object. The moon is a satellite of the Earth. There are also many man-made satellites.
Shooting star: another name for a meteor. Meteors look like fast-moving stars in the sky.
Solar flare: a sudden burst of hydrogen gas from part of the sun's surface.
Solstice: the summer solstice is the longest day of the year, and the winter solstice is the shortest day of the year.

Tidal surge: an extremely high tide that occurs when atmospheric pressure is unusually low.
Tsunami: a powerful wave produced by earthquakes, volcanic eruptions, or landslides beneath the sea.

Universe: a name used to refer to everything that exists, which includes all the millions of stars, planets, and moons, as well as all energy and space.

Vapor: another name for steam, a cloud of tiny liquid drops.
Wanes: when the moon appears to be becoming smaller. The moon wanes until it reaches the new moon phase, when the moon does not appear in the sky.
Waxes: when the moon appears to be growing larger. The moon waxes until it becomes a full moon.

Year: the time length of about 365 days for the Earth to make its journey around the sun.

Zodiac: the groups of stars which the Earth seems to travel within are called the signs of the Zodiac. Astrologers read the patterns in an attempt to find out what the future holds for people.

71

■ Ways of speaking

The sun
"A place in the sun" is a good place to be, whether it is really in the sunshine or where people like and respect you.

"Make hay while the sun shines" means taking the opportunity to do something while you can. Do you look on "the sunny side of things?"

The stars
To "sleep under the stars" is to sleep outdoors.

To "thank your lucky stars" is to say you are grateful for a piece of good luck that came your way.

The moon
The Latin word for the moon is *luna*. "Lunatic" (or "loony" for short) means "moonstruck" because the moon was once thought to cause insanity.

A "blue moon" is the second full moon appearing in the same month. This seldom occurs, so when we say something happens "once in a blue moon," we mean it happens very rarely.

Day and night
Something elegant and lovely to look at is said to be "as beautiful as the day is long."

"It's as clear as daylight" means that you completely understand something.

"They are like day and night" means two people or things are complete opposites.

The seasons
"The spring of your life" is when you are young, while "the autumn of your life" is past middle age.

The Earth
"To come down to Earth" means to stop daydreaming and get on with doing practical things.

The sky and the heaven
Saying "thank heavens!" shows that you are glad and relieved about something.

To "move heaven and Earth" to get something done means that you do everything you possibly can to make sure it happens.

Someone in "seventh heaven" is very happy.

"When the sky falls" means there is a terrible disaster.

"The sky's the limit" means you have the ability to do just about anything, and there is nothing to stop you from doing it.

Volcanoes
"Sleeping under a volcano" means taking a big risk. Everything seems safe now, but something awful could happen at any moment.

■ Here is a list of museums you can visit to learn more. Stop by your library to learn of museums in your area.

The Discovery Museum, Inc.
4450 Park Avenue
Bridgeport, CT 06604

Museum of Astrogeology, Meteor Crater
603 North Beaver, Suite C
Flagstaff, AZ 86001

Smithsonian Institution
1000 Jefferson Drive S.W.
Washington, D.C. 20560

Space Center Houston
1601 NASA Road One
Houston, TX 77058

The Science Museum of Minnesota
30 East 10th Street
St. Paul, MN 55101

California Academy of Sciences
Golden Gate Park
San Francisco, CA 94118

INDEX

The entries in **bold** refer to whole chapters on the subject.

A
Air, A cloak of air surrounds the Earth, 26, The effects of the atmosphere on Earth, 27, The wind is a great, flowing current of air, 28; *see also* 70
Alps, 45, 46, 59
Altiplano, 47
Altitude, 28
Andes, 46, 47, 58, 63
Andromeda galaxy, 16
Anemometer, 36
Antarctica, 23, 42, 58, 65, 70
Arctic, 23, 70
Arctic Ocean, 38, 58
Astrologer, 13, 70
Astronaut, 7, 26, 64
Astronomy, 4, 10, 17, 68, 70
Atmosphere, 5, 11, 22, 26–7, 36, 37, 70

B
Barometer, 36, 37
Bay, 56, 70
Beaufort scale, 29
Big Dipper, 13

C
Carbon dioxide, 26, 70
Caves, A hidden world under the ground, 54, Speleologists explore and study underground caves, 55
Climate, 21, 24–5, 30, 35, 40, 59, 70
Clouds, 24, 26, 28, 29–34, 35, 36, 37, 48, 67
Coast, 29, 38, 41, 45, 58
Comet, 4, 5, 10, 70
Condensation, 31, 70
Constellation, 12, 13
Continents, 24, 38, 46–48, 59, 70
Copernicus, Nicolaus, 4, 68
Core of the Earth, 44, 70
Crater, 7, 48, 49
Crevasse, 35
Crust of the Earth, 44, 45, 48, 52, 62, 70
Currents, 40

D
DaVinci, Leonardo, 69
Day, The sun's daily journey across the sky, 18, The continuous cycle of day and night, 19
Delta, 43
Deserts, 24, 57

E
Earthquake, Earthquakes happen when the Earth's crust moves, 52–53; *see also* 41, 59
Eclipse, of the moon, 9, 64, 70, of the sun, 15
Einstein, Albert, 68
Epicenter, 53, 70
Equator, The seasons at the poles and the equator, 22; *see also* 24–25, 70
Equinox, 20, 21, 70
Erosion, 45, 55–56, 61, 70
Estuary, 43, 65
Evaporation, 30, 70

F
Fjord, 43
Franklin, Benjamin, 62
Frost, The endless work of water, wind, and frost, 56–57

G
Galaxy, The Milky Way is like an island among the billions of galaxies in the universe, 16–17; *see also* 70
Galilei, Galileo, 64, 68
Glacier, 35, 42–7, 57, 59, 63, 65
Gravity, 17, 26, 40, 70
Greenwich meridian, 70
Gulf Stream, 40

H
Halley's comet, 10
Helium gas, 12, 37
Hemisphere, 70, Northern, 12, 13, 20, 62, Southern, 20, 64
Hillary, Sir Edmund, 64
Himalayas, 45, 47, 59
Hurricane, 29
Hydrogen, 12, 14, 70
Hygrometer, 37

I
Ice, Snow is water transformed into ice, 34; *see also* 56–7, 62
Ice Age, 43, 70
Icebergs, Pack ice and icebergs, 42; *see also* 70
Incas, 47

J
Jupiter, 4, 64, 68

K
K2, 44, 64
Kepler, Johannes, 68
Kon-Tiki, 40

L
Lagoon, 43
Landslides, 56
Lava, 48–50
Legends, 63, 71
Lightning, 34, 64

M
Magellan, 64
Magma, 48–50
Mantle of the Earth, 44–5, 48, 52, 62, 71
Map of the Earth today, 58–59
Mars, 4, 5
Mercury, 4, 5, 64
Meteor, 11, 71
Meteor Crater, Arizona, 11
Meteorites, 7, 11, 64, 71
Midnight sun, 22
Milky Way galaxy, 16, 17, 71
Moisture, 34, 35, 71
Month, 8, 9, 15, 71
Moon, Earth's companion in space, 6, Without the sun we would never see the moon, 7, The Earth and the moon play hide and seek, 8, The moon waxes and wanes each month, 9; *see also* 15, 31, 40, 62–63, 68, 71, 72
Moraine, 57
Mountains, Mountains have the power to frighten and fascinate, 44, Earth's mighty internal energy folds and shatters rocks, 45, Mountains form chains stretching across continents, 46, The Himalayas and the Andes: mountain fortresses, 47; *see also* 26, 29, 30, 35, 38, 42, 49, 53, 56–59, 62–63
Mount Everest, 26, 44, 47, 64

N
Neptune, 4
Newton, Isaac, 64, 68
Night, The continuous cycle of day and night, 19, Auroras illuminate the polar night sky, 23
Nitrogen, 26

O
Oceans, 29–30, 38–40, 42–43, 53, 62, 65, 71
Orbit, 4, 11, 17, 22, 71
Oxygen, 26, 71

P
Pacific Ocean, 38, 40, 46, 53, 58
Pack ice, 42, 71
Phases of the moon, 8, 71
Planets, 4–5
Plates of the Earth, 45, 52, 71
Pluto, 4–5
Poles, The seasons at the poles and the equator, 22, Auroras illuminate the polar night sky, 23, Deserts of baking sand and freezing ice, 24; *see also* 38, 42, 62, 71
Ptolemy, 68

R
Radar, 37, 71
Rain, As the clouds grow, it begins to rain, 30, Clouds make landscapes in the sky, 31, The endless work of water, wind, and frost, 56; *see also* 24, 25, 37, 45–46, 49, 54, 57, 69
Rainbows, are bridges of colored light across the sky, 33
Rain forest, 25
Richter scale, 53, 71
Rivers, 30, 35, 43, 54–56, 65
Rotation, 4

S
Satellites, Using satellites and radar for weather forecasting, 37; *see also* 11, 17, 37, 71
Saturn, 4–5, 64
Sea currents, 40–41
Seas, Our blue planet, 38, The kingdom of the deep, 39, The ocean waters are always moving, 40–41; *see also* 42, 43, 64
Seasons, The four seasons, 20, Spring, summer, autumn and winter, 21, The seasons at the poles and the equator, 22; *see also* 25, 62, 72
Serac, 35
Shooting star, 11, 71
Snow, is water transformed into ice crystals, 34, The formation of glaciers, 35; *see also* 24, 25, 29, 46, 47, 49, 57
Solar flare, 14, 71
Solar system, The Earth circles the sun, 4, The Earth is part of the solar system, 5; *see also* 6, 10, 11, 16, 18
Solstice, 20, 21, 71
Speleologists, explore underground caves, 55, Stalactites and stalagmites, 54
Stars, The birth of stars, 12, The stars have long been a map for travelers, 13, Our star, the sun 14; *see also*, 4, 5, 16, 17, 27, 72
Sun, Without the sun we would never see the moon, 7, Our star, the sun, 14, The sun hides behind the moon in a solar eclipse, 15, The sun's daily journey across the sky, 18; *see also* 4–6, 9–10, 12, 19–23, 26–27, 30–31, 33, 40, 62–64, 66, 68, 72
Sunrise and sunset, 19, 27, 66

T
Telescope, 12, 14, 17, 64, 68
Thunder, Lightning strikes and thunder rumbles during a thunderstorm. 32; *see also* 29, 31, 62–63, 67
Tidal surge, 43, 73
Tides, 40, 41, 43
Time and time zones, 18, 62, 66
Trifid Nebula, 12, 16
Tropics, 25, 29, 43

U
Universe, 4–18, 68, 71
Uranus, 4–5, 63

V
Vapor, 30–31, 71
Venus, 4–5
Verne, Jules, 69
Volcanoes, are mountains that breathe fire, 48, The life cycle of volcanoes, 49, In search of the secrets of volcanoes, 50, There are thousands of active volcanoes on Earth, 51; *see also* 38, 47, 57, 65, 72

W
Water, The endless work of water, wind, and frost, 56
Water cycle, 30
Waves, 41, 56
Weather, How is a weather forecast made, 36, Using satellites and radar for weather forecasting, 37; *see also* 67
Wind, is a great, flowing current of air, 28, Wind can produce a soothing breeze or severe damage, 29, The endless work of water, wind, and frost, 56

Y
Year, 4, 18, 71

Z
Zodiac, 13, 71